NUTRITION AND HEALTH ADVERTISING:

EVIDENCE FROM FOOD ADVERTISING 1977 - 1997

NUTRITION AND HEALTH ADVERTISING:
EVIDENCE FROM FOOD ADVERTISING 1977 - 1997

PAULINE M. IPPOLITO
JANIS K. PAPPALARDO

Novinka Books
New York

Library of Congress Cataloging-in-Publication Data

Ippolito, Pauline M.
Nutrition and health advertising: evidence from food advertising, 1977 – 1997 / Pauline M. Ippolito and Janis K. Pappalardo.
p. cm.
Includes bibliographical references and index.
ISBN: 1-59033-734-4.
1. Advertising—Food—United States. 2. Food—Labeling—United States. 3. Nutrition policy—United States. 4. Deceptive advertising—Law and legislation—United States. 5. Health behavior—United States. [DNLM: 1. Advertising—trends—United States. 2. Food labeling—trends—United States. 3. Advertising—economics—United States. 4. Advertising—legislation & jurisprudence—United States. 5. Nutritive Value—United States. WA695 I645n 2003] I. Pappalardo, Janis K. II. Title.

HF6161.F616I57 2003
659.1'9664'00973—dc21

2003008970

Nova Science Publishers, Inc.
400 Oser Ave, Suite 1600
Hauppauge, New York 11788-3619
Tele. 631-231-7269 Fax 631-231-8175
e-mail: Novascience@earthlink.net
Web Site: http://www.novapublishers.com

Printed in the United States of America

ACKNOWLEDGMENTS

Such a large scale data project could not have been completed without the support of many individuals. We are particularly grateful to our coders, who quickly and conscientiously absorbed a complex coding system and patiently persisted through the coding of hundreds of advertisements. Our sincere thanks go to Andrew W. Blume, Tiffany E. Cochran, Brandi D. Hackworth Day, Mark N. Hwang, Sean P. Logue, Maren A. Mikkelsen, Nicole L. Piekarski, Victoria R. Roubal, Merritt A. Thomas, and St. John P. Wiles, our talented coders, and most especially to Elizabeth A. Autry, who helped to train, coach, administer, and verify the coding of our advertisements through the entire data portion of the project.

We are also very grateful to John H. White, our programmer, who developed the computerized version of the coding instrument, the automated verification system, and the data base design. We also thank Debra Jones Ringold of Willamette University, who graciously reviewed the coding form and technique, and provided advice on improving the design.

We also owe a debt of gratitude to numerous librarians in the Washington, D. C. area, who helped us track down magazines for the study. Most especially, we thank Marcia Dysart at the Periodicals Department at the Enoch Pratt Free Library in Baltimore, and Joseph Puccio at the Library of Congress, who were of great help to us. We thank Helen W. Small of our secretarial staff, who assisted in the copying of so many advertisements. Also, we are grateful to Sandy Lin, a research analyst with the Bureau, who carefully read the entire manuscript for editorial consistency. The report is much improved for her efforts.

We are also particularly grateful to Paul A. Pautler, who supported the idea from the beginning, generously provided advice at all stages of the project, and commented on various drafts as the project progressed.

Finally, we would be remiss in not explicitly recognizing the substantial support of the Federal Trade Commission. It is a testament to the agency that it has a long history of supporting data collection and analyses relevant to its missions, designed to evaluate and inform its policy decisions.

Pauline M. Ippolito
Janis K. Pappalardo

CONTENTS

* This report was augmented by a subject index compiled by the publisher.

List of Figures

LIST OF TABLES

SUMMARY[*]

This study examines new data on the types of claims made in food advertising during the years 1977 to 1997. The study's primary focus is the use of nutrition-related claims. Besides providing a wealth of data on the basic content of food advertising over time, we have two additional goals: first, to better understand the economic forces affecting the flow of nutrition information to consumers in marketing, and second, to examine firms' incentives to focus on nutrition in advertising under the various policies adopted during these years, including those adopted after the *Nutrition Labeling and Education Act of 1990* (NLEA).

In the latter half of the twentieth century, dietary research has focused on the role of diet in the major chronic diseases, including heart disease and cancer. Since consumers choose their own diets, this has lead to a debate about when and how to bring this growing body of knowledge to consumers, and of particular relevance for this study, what role food marketing might play through claims about nutrients, diet, and health.

As this debate played out in policy circles, the regulation and enforcement policies governing nutrition-related claims in advertising and labeling changed several times, culminating in the current post-NLEA environment. These regulatory shifts provide the opportunity to test various hypotheses about firm behavior under different enforcement policies. In a series of earlier studies, we examined consumers' dietary choices during the different regimes, but with one exception, our earlier efforts contained no data on the claims actually made in marketing, only information about the policies governing those claims. This study attempts to fill this void by creating and analyzing a large, systematic database of advertising content for the years 1977 to 1997.

METHODOLOGY AND ADVERTISING SAMPLE

Television is the medium used most intensively for food advertising, but unfortunately, no archives exist that allow us to create a systematic sample for study. Magazine advertising is the second largest category of food advertising. We compile a large, systematic sample of food advertising from 5 of the leading women's magazines and 3 of the most popular general readership magazines. Claims are extracted from the advertisements using state-of-the-art techniques for reliability, as described in Chapter 2. The sample has 11,647 food advertisements.

---

[*] The authors are economists with the Bureau of Economics, Federal Trade Commission. The views expressed in this study are those of the authors and do not necessarily represent the view of the Federal Trade Commission or any individual Commissioner.

Magazines
Better Homes and Gardens
Good Housekeeping
Ladies Home Journal
McCall's
Women's Day
Reader's Digest
Newsweek
Time
Months
February
June
October
Years
1977 - 1997

To our knowledge, these data provide the most comprehensive examination of magazine advertising in a particular market ever undertaken. We believe that the data developed for this project present a very accurate and complete picture of the types of nutrition-related claims made in magazine food advertising over the years 1977 to 1997.

BROAD TRENDS IN FOOD ADVERTISING

Magazine Advertising Grows Relative to TV; Number of Ads Falls

As shown in Chapter 3, the price of advertising to 1000 households for both television and magazines has grown faster than other producer prices since the early 1980s, and the price of television advertising has grown relative to magazines. The proportion of food advertising dollars spent in magazines has increased relative to television during the same period, from approximately 9 percent of total spending in the late 1970s to 13.6 percent in 1997. The number of food advertisements in our sample has fallen consistently since the mid-1980s, from approximately 600 ads in 1977 and in 1986, to 400 ads in 1997, a reduction that matches trends in comparable industry data.

Nonnutritive Claims Are Common in Food Advertising; Most of These Claims Fall Over Time

Claims about food characteristics other than nutrients have always been prominent in food advertising. We collect information on claims about taste, aroma, texture, convenience, whether the product is new or improved, its varieties, suggestions for use, price or coupons, and promotional offers.

In each of these categories, except the *new/improved* category, the percent of ads with claims trends downward. The largest trend is for the *taste/aroma/texture* category, where claims are steady at approximately 85 percent of food ads until the mid-1980s, but then fall by about 25 percent to 67 percent of ads by 1997. *New/improved* claims are the only category with a significant upward trend, and this is the category that might be related to the development of nutritionally improved products. Approximately 15 percent of ads in 1977 have a *new/improved* claim, compared to more than 25 percent in 1984 and 1997.

NUTRIENT CONTENT CLAIMS IN FOOD ADS

Nutrient content claims are statements or terms referring to a specific nutritional characteristic of a food, *e.g., low fat, more fiber,* or *contains vitamin E.* As described in Chapter 4, the study collects data for all the major nutrients, as well as other miscellaneous specific nutrition-related claims. For each nutrient, claims are coded in two subcategories, *level claims*, that describe the absolute amount of a nutrient, such as *low fat* or *high fiber*, and *comparative claims*, that compare the amount of a nutrient in a food to something else (even if unstated), such as *less fat* or *more fiber.*

Fat Claims

Fat claims include all claims about unspecified types of fat. This category does not include claims about specific types of fat, such as saturated fat, or other specific fat claims, such as, *made with canola oil*, which are coded separately. Both *fat level* and *fat comparative* claims are included in the overall *fat* claim category, and an ad can have both types of claims.

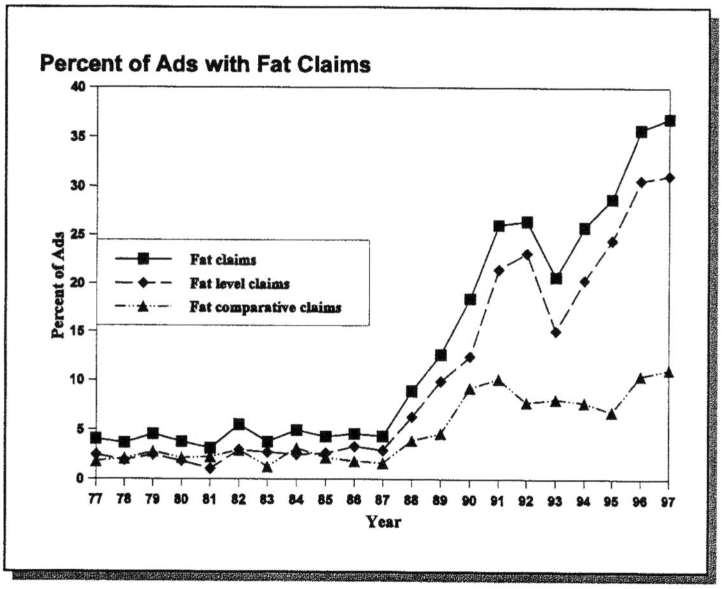

In 1997 fat claims are the most frequent nutrient content claim by far.

As shown in the graphic, the dominance of fat claims is a relatively recent phenomenon; fat claims are made in less than 5 percent of ads before 1987. Comparative fat claims grow in parallel to level claims until 1990, when approximately 10 percent of ads use them and where they remain in 1997.

Saturated Fat and Cholesterol Claims

Saturated fat and cholesterol are the lipids most clearly identified with the risk of heart disease. The pattern of use of saturated fat and cholesterol claims over time is distinctly different than for fat claims. Both are used increasingly through 1990, before falling substantially after 1990. Comparative claims follow the same pattern and are essentially eliminated by 1997.

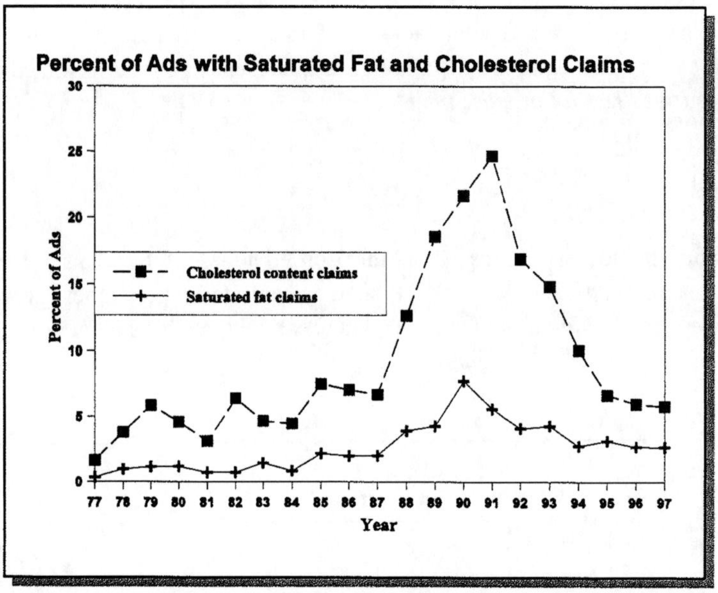

Saturated fat and cholesterol claims fall after 1990. Focus shifts to total fat.

Other Nutrient Content Claims

Advertising for other major nutrients, such as calcium, fiber, and sodium, are described in Chapter 4. Most generally follow the pattern of rising prior to 1990 and falling or remaining relatively stable in the post-1990 period. Comparative claims generally rise prior to 1990 and fall after 1990 to very low levels.

Calorie and Dieting Claims

Claims about calories or weight control, including *diet* claims, are a significant feature of food advertising throughout the period. Calorie claims are approximately evenly split between level and comparative claims until the early 1990s, when comparative claims fall faster than level claims.

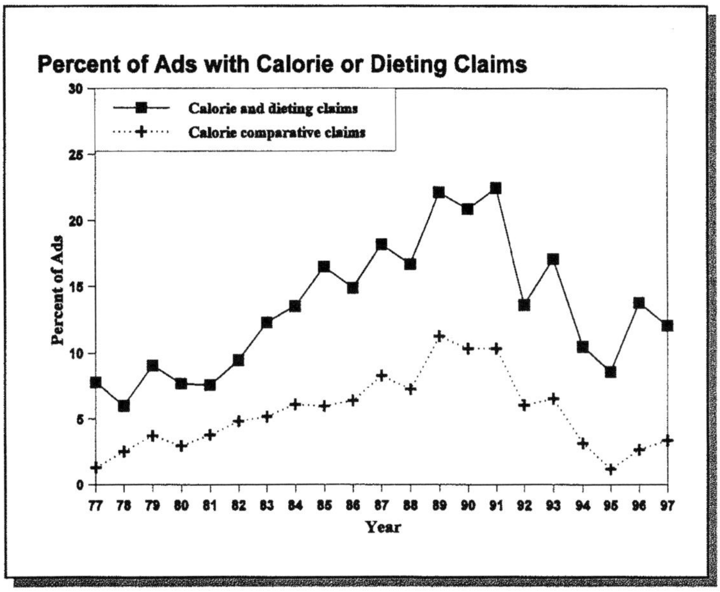

General Nutrition Claims

General nutrition claims are statements or terms, other than nutrient content claims or health claims, that indicate a potential health or nutrient advantage of an advertised food.

General nutrition claims are quite common in food advertising. In 1977, 50 percent of all ads have a general nutrition claim. Their use rises to nearly 70 percent of ads by 1983 and is steady through 1990, before falling back to 56 percent of advertisements in 1997. Data on subcategories of claims are described in Chapter 4.

General nutrition claims are more common than specific nutrient or health claims throughout the period, but the gap between them narrows dramatically over time. In the broadest sense, the data indicate a sustained movement towards greater use of specific nutrition claims in place of, or in addition to, general nutrition claims during the years 1977 to 1997.

General Nutrition Claim Subcategories
health/healthy
smart/right choice
good/better for you
nutritious/nutrients
wholesome
enriched/fortified
light/lighter
lean/leaner
guilt free/no guilt/cheating
fresh
energy
natural/no artificial/real/pure
youth/fitness/well-being
other general nutrition terms

HEALTH CLAIMS IN FOOD ADVERTISING

Health claims are statements about *specific health effects* of nutrients or foods. Within health claims, we focus on three subcategories of claims: *disease claims*, which explicitly refer to a disease; *affiliated claims*, which refer to conditions closely affiliated with disease, namely, serum cholesterol levels, high blood pressure, and heart claims that are not specific to disease, as in *heart smart*; and *other nondisease health claims*, which are health claims that do not fit in either of the previous categories. These other nondisease health claims, such as *builds strong bones*, would often be considered structure-function claims in FDA terminology. Note that serum cholesterol claims do not include cholesterol content claims, such as *no cholesterol.*

Explicit disease claims are not the majority of health claims during this period. Disease claims are made in less than one percent of ads prior to 1984, and in less than 4.6 percent of ads per year throughout.

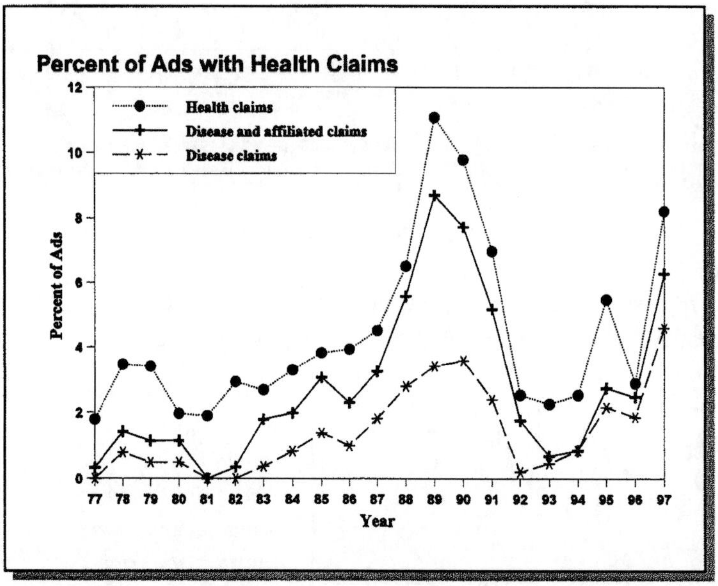

Disease and affiliated claims are the majority of health claims. These claims peak in 1989.

When affiliated claims are considered with disease claims, the picture changes. Disease and affiliated claims do constitute the majority of health claims from 1983 to 1992, and again after 1995. The percentage of ads with a disease or affiliated claim is well under 2 percent through 1982 and peaks at 8.7 percent in 1989. The use of disease and affiliated claims falls precipitously after 1990 and begins rising again only in 1995. By 1997, 6.3 percent of advertisements include a disease or affiliated claim, 72 percent of the 1989 peak.

In the early years of the sample, most health claims are other nondisease health claims, often dealing with bones, teeth, digestibility, or regularity. Similarly, in the early 1990s, when disease and affiliated claim use is very low, use of other nondisease health claims grows.

When explicit disease and affiliated claims are not used, producers appear to shift to less explicit health claims where possible.

Heart and Serum Cholesterol Claims

Heart-related claims are the most common health claims by far. Heart or serum cholesterol claims are used a bit in the late 1970s, and then begin again in 1983, rising substantially to a peak use of 8.2 percent of all ads in 1989, before falling dramatically in the early 1990s. In 1997, 3.4 percent of ads include a heart or serum cholesterol claim, 41 percent of the peak use.

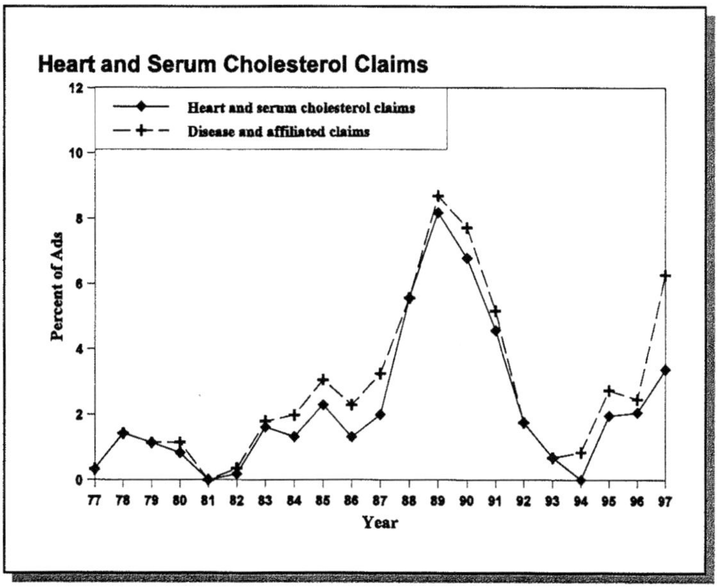

Heart and serum cholesterol claims are the most common health claims by far.

Cancer Claims

Cancer claims are much less frequent than heart claims throughout the years 1977 to 1997. Cancer claims essentially begin in 1984 highlighting fiber content for cereals. Fruit and juice producers joined the cereal producers in the 1980s, but cancer ads never rise above one percent of all food ads during this period. Cancer claims begin again in 1994 and rise to 2 percent of ads in 1997. The post-1990 claims are primarily from juice producers, with cereal producers joining again in 1997 following the FDA approval of a new oat-heart claim that triggered increased health claim competition among cereals.

Other Health Claims

The evidence indicates that other health claims are used less frequently. Chapter 5 presents evidence on claims dealing with osteoporosis and bones, hypertension, birth defects, diabetes, cell damage, oxidization, free radicals, tooth decay, and regularity, as well as a residual category of all other health claims.

REGULATION AND ADVERTISING CLAIMS

Nutrition-related claims have been the subject of considerable regulatory and enforcement scrutiny during the years 1977 to 1997. Advertising claims are under the primary jurisdiction of the Federal Trade Commission (FTC) and food label claims are regulated by the Food and Drug Administration (FDA). Both agencies initiate major rulemakings during the years of our sample. The study examines the timing of changes in the use of nutrition and health claims relative to key regulatory and enforcement events.

Key Regulatory Events

For the statistical analysis, we focus on five key events:

- **FTC Food Rule Decisions: April 1980 and December 1982** The first two events are associated with the FTC's Food Rulemaking of the late 1970s and early 1980s. The first event occurs in April 1980, when the FTC ends Part II of the Food Rule, which would have regulated general nutrition claims, such as *health food* claims, and emphatic nutrition claims, such as `lots of fiber`. The Commission also directs the staff to continue with an effort to define conditions for fatty acid and calorie claims, heart-related health claims, and some other nutrient and general claims. On December 17, 1982, the Commission votes to end the remaining portions of the Food Rule, opting instead to proceed on a case-by-case basis under its general deception authority. Thus, by early 1983 it is clear that nondeceptive claims about nutrition issues, including explicit health claims, will be considered favorably by the FTC. It is in this environment that Kellogg initiates planning for its fiber-cancer advertising campaign that first airs in October 1984.
- **FDA Health Claim Proposal: August 1987** Health claims also raise the risk of legal action at the FDA, which prior to 1987 essentially bans all diet-disease claims for foods. After much public discussion, in August 1987 the FDA proposes a rule that would allow nondeceptive health claims on labels under a less restrictive standard. This proposal is widely viewed as reducing firms' legal risk in making certain health claims.
- **FDA Rescinds 1987 Proposal: February 1990** After considerable public debate, FDA rescinds the 1987 proposal in February 1990. This is followed in July 1990 by publication of a more restrictive FDA proposal for food claims, and in November 1990 by the *Nutrition Labeling and Education Act of 1990* (NLEA), legislation which lays out standards for revising food labeling rules. The events of 1990 are

broadly perceived to restrict producers' use of health and other nutrition claims and to set the stage for a revision of labeling rules under the NLEA.

- **Final NLEA Rules Effective; FTC Food Policy Statement: May 1994** Following the enactment of the NLEA, the FDA develops extensive regulations covering all aspects of the food label. This is a period of considerable uncertainty as rules are proposed and finalized. The major proposal is issued in November 1991. Label regulations governing health claims are effective in May 1993 and nutrition claims in May 1994. Also in May 1994, the FTC issues a policy statement harmonizing advertising policy with the new food labeling rules. In December 1995, the FDA also issues a proposed rule to clarify key features of NLEA regulations, but this proposal has not been finalized.

Key features of the NLEA-based rules include a listing of approved nutrition claims, a prohibition of unapproved nutrition claims, explicit requirements for nutrient content claims, triggered disclosures in some cases, *e.g.,* for comparative claims, and provisions for a limited number of health claims with specific restrictions on which foods can make such claims.

Health Claims and Regulations

The policy changes during these years are most pronounced for health claims, especially disease and affiliated claims. This study uses linear and probit regression techniques to examine whether disease and affiliated claims increase or decrease following key regulatory events. Among the findings are the following:

- **Health Claim Use Changes With FTC Food Rule Decisions** Following the 1980 FTC decision directing the staff to draw up explicit regulations for heart-health claims, the low level of health claims in use at the time falls to near zero. Conversely, the 1982 FTC decision to return to a case-by-case approach for health claims is followed by a statistically significant increase in the use of disease and affiliated claims to approximately 2 percent of ads.
- **Health Claims Increase Significantly Following the 1987 FDA Proposal** The FDA's August 1987 proposal to allow health claims is followed by a statistically significant increase in the use of health claims in advertising. Health claims increase by 5 percentage points from a base of approximately 2 percent of ads. This evidence is consistent with the view that the FDA label rules have an important influence on producers' willingness to make advertising claims.
- **FDA Reversal and Other 1990 Events Are Followed by Large, Statistically Significant Drop in Health Claims** In the period following February 1990, when FDA reverses its 1987 proposal, health claims in advertising fall rapidly to low levels. The size of the drop is sufficient to eliminate the increase following the 1987 proposal. These results are highly statistically significant in both linear and probit specifications. Again this evidence is consistent with the view that FDA labeling rules affect the claims producers are willing to make in advertising.

- **Health Claims Rise Again in the Post-NLEA Environment, But Not to Previous Levels** After 1994, when the FDA's NLEA-based rules are effective and the FTC has issued its harmonization statement, health claims again increase. The growth in disease and affiliated claims comes mostly in the last two years of the post-1994 period. This pattern suggests that the FDA's December 1995 proposal to simplify the rules for health claims may have been important to advertisers. This proposal, which has never been finalized, makes it clear that the long and rather complicated model statements in the original NLEA health claim regulations are not required and proposes other simplifications in the rules.

Evidence indicates that advertisers respond significantly to regulatory rules for health claims.

Nutrient Claims and Regulation

Regulatory events could also affect the use of nutrient content claims, both directly because the rules govern nutrient claims, and indirectly because nutrient claims are often used with health claims or may be spurred by the increased focus on diet-health issues engendered by those claims.

We focus on the results for 8 primary nutritional characteristics of foods: total fat, saturated fat, cholesterol, sodium, fiber, calcium, vitamins/minerals, and calorie/diet claims. Among the findings are the following:

- **1980 End of Part II of the FTC's Food Rule Is Not Followed by Much Systematic Change in the Use of Nutrient Claims** Only 3 of the 8 nutrients have significant movements after this event, indicating only limited change. Fiber and sodium claims increase significantly.
- **1982 End of the FTC Rulemaking and 1987 FDA Proposal Are Both Followed by Systematic Increases in Nutrient Content Claims** Both of these events relax the policy towards health claims, and the 1982 event affects some nutrient claims directly. Both events are followed by systematic changes in the use of nutrient claims. Significant changes occur for 5 out of 8 nutrients after the first event, and for 6 out of 8 nutrients after the second event. All of the significant changes are positive, indicating a systematic increase in nutrient claims for most nutrients after these events.
- **After 1990 and 1994, Growth in Nutrient Content Claims Slows and Then Drops** After 1990, 5 of the 8 nutrients show significant changes, but only 3 of the 5 increase. After 1994, when the NLEA rules are final, 6 of the 8 nutrients have significant changes, but only 2 of the 6 increase. Fat, and to a lesser extent calcium, are the two nutrients where content claims continue to grow in the post-1994 period. In contrast, producers reduce their focus on saturated fat, cholesterol, sodium, and calories after the NLEA rules in 1994. These changes are all statistically significant.
- **Comparative Claims Rise Prior to the NLEA Rules and Fall After the NLEA Rules** Comparative claims are more restricted under the NLEA rules and must include several triggered disclosures. Prior to the NLEA, the use of comparative

claims increases significantly for 5 of 8 nutrients after 1982 and for 6 of 8 nutrients after the health claimpolicy change in 1987. After the NLEA, use of comparative claims changes significantly for 5 of 8 nutrients after the 1990 event, and for 6 of 8 nutrients following the 1994 events, but only 3 of these 11 significant changes are increases. Most notably, when the NLEA rules are final in 1994, comparative claims fall for 7 of 8 nutrients (6 significant). The only exception is for total fat, which exhibits no significant change.

One of the most consistent changes in the post-1994 period is the systematic movement away from comparative claims for all major nutrients except total fat.

General Nutrition Claims and Regulation

General nutrition claims, such as *healthy* or *nutritious*, are also potentially affected by the regulatory events of this period. Some general terms, such as *healthy*, are directly regulated, and more broadly, these general claims could complement or substitute for specific claims subject to the rules.

Statistical results are presented in Chapter 6. The use of general health claims seems to fall when specific claims increase and to rise when specific claims are more restricted, suggesting that general claims substitute somewhat for more specific claims when those are restricted.

Health Claims Across Food Groups

We would expect the changing regulations to affect advertising in some food groups more than others. Some foods have a larger role to play in improving diets. Moreover, under the rules implementing the NLEA, health claims are limited to foods that are "best" on the dimensions relevant to the particular health claim, "not bad" on other key dimensions, and "nutritious" in the sense that they provide a minimum level of nutrition on at least one of six specified nutrients. By limiting health claims to these particular foods, it is hoped that producers will find it more profitable to promote these foods, and that as a result consumers will be more successful in improving their diets. If these presumptions are correct, the NLEA rules should increase health claims for these foods, increase advertising for them, and reduce the use of health claims by sellers of other foods.

These issues are examined in Chapter 6. Among our findings:

- **Following 1987 FDA Proposal, Health Claims Increase in the Cereal/Bread, Fats & Oils, Meat/Egg, and Poultry/Fish/Grain Categories** The largest increases in health claims occur after the 1987 proposal, with the percentage of ads with health claims increasing by 25.3 percentage points for Fats & Oils, 16.5 points for the Cereal/Bread category, 10.3 points for Meat/Eggs, and 2.0 points for Poultry/Fish/Grains.
- **Number of Fruit, Vegetable, and Juice Advertisements Drops Significantly After 1990; Only Orange Juice Ads Have Health Claims** The amount of

advertising in the Fruit/Vegetable/Juice category drops significantly in the post-NLEA period. In our sample, the category averages approximately 100 advertisements per year through 1990, when the number begins dropping, and stabilizes after 1993 at approximately 50 ads per year, half the pre-1990 level.

Advertising in the category drops sharply, but orange juice producers continue to use health claims.

Those producers who continue to advertise are more likely to use health claims. But with one exception, the only health claims in the category after 1990 are from orange juice producers.

- **After 1990 Health Claims Increase for Dairy; Decrease for Fats & Oils, Meats/Eggs, and Poultry/Fish/Grains** In the post-NLEA period the percentage of advertising with health claims increases significantly for the Dairy category, which grows by 5.1 percentage points. More sizable effects are found in the food categories where health claims fall. The percentage of advertising with a health claim falls by 43.7 percentage points for Fats & Oils, by 10.4 points for Bread/Cereals, by 7.4 points for Meat/Eggs, and by 2.5 points for Poultry/Fish/Grains.
- **Advertising Does Not Increase in Any Food Category in the Post-NLEA Years** Regressions relating the number of advertisements per month to the key regulatory events show remarkable stability prior to 1987. After the 1987 proposal, the only category with a statistically significant change is Desserts/Snacks, where the number of ads per month drops by 34 percent. In the post-1990 period, advertising falls for 8 of the 9 food groups, with significant reductions for Cereal/Bread, Fruit/Vegetables/Juice, and Fats & Oils. After the final NLEA rules in 1994, advertising falls for 6 of the 9 food groups, with Fats & Oils and Fruit/Vegetables/Juice experiencing further significant declines. The 3 food groups where advertising increases after 1994 all reflect a return to the level of advertising in 1990.

Evidence shows no increased advertising in "good food" categories in the post-NLEA period but reduced advertising in other select categories.

- **Health Claims Not Used for Desserts/Snacks or Soft Drinks Before or After the NLEA** Some provisions in the NLEA rules are motivated by a concern that producers of empty or otherwise nutritionally deficient foods would use health claims in marketing. In fact, the requirement that foods must have certain nutrition value to qualify to make a health claim is commonly called the "jelly bean rule," reflecting the fact that without the requirement, an advertiser of jelly beans could legally make a heart claim under NLEA rules (jelly beans are low in fat and saturated fat and contain no cholesterol).

To explore the magnitude of this perceived problem, we examine two food categories in detail: Drinks, which includes all carbonated soft drinks and all fruit-flavored beverages (but not juice or milk), and Desserts/Snacks, which includes desserts, sweets, donuts, salty snacks, and related items. The evidence indicates that with a few trivial exceptions, health claims are never used in marketing foods from either of these categories. The amount of advertising falls in both categories over time, but these declines precede the NLEA. Thus, the evidence provides no support for the view that health claims for "junk foods" is a significant concern during these years.

Summary of Findings on Regulations

Overall, the evidence is consistent with the view that the content of food advertising varies considerably with changes in regulation and enforcement. The use of health claims varies most, as expected given the significant changes in policy towards these claims. But nutrition claims also vary a great deal following these events, as competition on health issues increases or decreases. Under the NLEA rules, the focus in advertising has shifted primarily to total fat away from saturated fat, cholesterol, calories, and other nutrients. Also, in the post-NLEA years, producers have moved away from comparative claims for all nutrients except total fat. The reasons for these results and their effect on consumer diets are important areas for further research.

ECONOMICS OF ADVERTISING: ISSUES AND EVIDENCE

Advertising is a major feature of consumer good markets. Firms have an incentive to try to draw consumers to consider their products, especially consumers who will become regular customers. By highlighting product characteristics in advertising, firms can attract consumers who value those characteristics, and if satisfied with the product, they are more likely to become regular customers. This simple mechanism underlies the information theory of advertising.

Specific Claims in Advertising

The economics literature contains considerable evidence that the introduction of advertising into markets can have a positive effect on market performance, through lower prices, product improvements, or beneficial changes in consumer purchases, for instance. Presumably because of the difficulties of acquiring data on the content of advertising, there is surprisingly little direct evidence on the information content of advertising and the economic forces that shape it. As a result, there is little evidence to judge whether advertising acts strictly as a signal of quality, a visible public expenditure, or through direct information provision.

These issues are explored in Chapter 7 of the report. Among our findings are the following:

- **Specific Nutrition Claims Have Become a Major Feature of Food Advertising** Specific nutrition claims are informative-type claims. A measure of their presence provides evidence on the information content of advertising in this dimension. We examine this issue in two ways: first, by determining the percentage of advertisements that include at least one specific nutrient content claim for any of 12 main nutrients, such as total fat, saturated fat, *etc*., and second, by determining the percentage of advertisements that have *any* specific nutrition-related claim recorded in our coding system. This second category includes the main nutrient content claims, as well as other specific nutrition-related claims, such as *made with canola oil*, *sugar free*, *etc.*

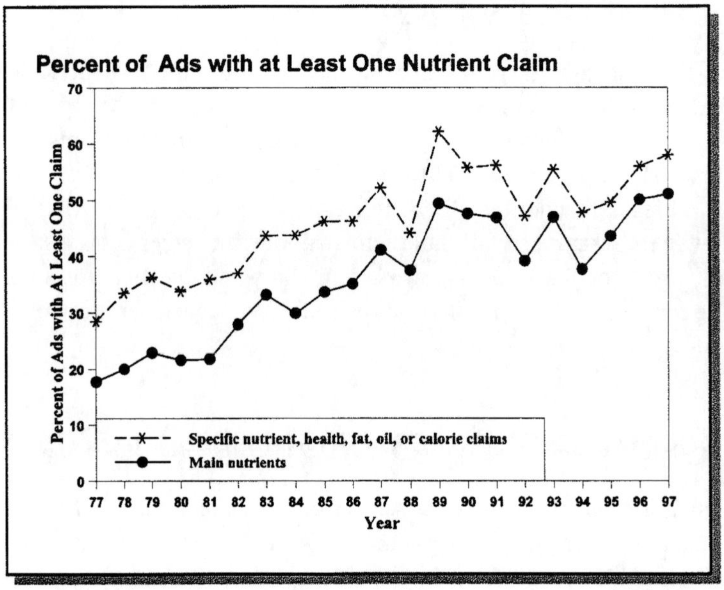

Both measures indicate substantial growth in the percentage of advertising with specific nutrition claims during the first half of our period. Since the late 1980s, however, the percentages have stabilized; approximately 40-50 percent of ads include claims about main nutrients and approximately 50-60 percent include claims from our broader class of specific nutrition-related claims. Despite changing policies and market conditions, approximately half of all food advertising since the late 1980s includes specific nutrition claims.

- **Other Specific Informative Claims Are Also Common in Food Advertising** Our data includes information on several other types of specific claims in food ads. Approximately 40 percent of ads include specific information for using the product, often by providing recipes that use the food. More than 50 percent of the ads include information about different varieties of the product, such as available flavors or package sizes. Approximately one-third of the ads make an explicit claim about the product's convenience for some use. Approximately 20 percent of the ads highlight that the product is new or has been improved. Finally, approximately 80 percent of the ads make a claim about the taste, texture, or aroma of the food.

Most food ads make multiple informative-type claims.

Taken together, this evidence illustrates that virtually all food advertisements in our sample make specific claims about the advertised product. In fact, most ads make multiple informative-type claims. Assuming that the nutrition label is credible to consumers, most of these claims involve *search* or *experience* characteristics, that is, characteristics that consumers can verify at purchase or after use.

Advertising and Unfolding: Does Competition Lead to Greater Information Disclosure?

One of the economic issues in advertising is the potential bias in the types of information provided by advertisers. Advertisers have an incentive to tell potential customers what is good about their product but not what is bad. This issue is of particular concern in multi-attribute products, such as foods, where claims about the desirable features could draw attention away from less desirable and unrevealed characteristics. Economic theory suggests that in many cases competition among producers can substantially reduce or eliminate this bias in the information provided by the market as a whole. This unfolding hypothesis holds that firms gaining sales by highlighting just one dimension will soon face competition from firms who point out superiority in other important dimensions as well.

The concern about incomplete information underlies some of the changes implemented in the NLEA rules. Under the NLEA rules, if producers make nutrient claims on their labels, they are required to highlight undesirable characteristics. Of course, triggered disclosures also reduce the incentive to make the original nutrient claims, because the claims are now more costly. We examine the unfolding hypothesis in several ways. Among our findings:

- ❑ **Mean Number of Lipids Featured in Ads Peaks in 1991; Falls 20 Percent After the NLEA** We have data on claims for 5 primary lipid characteristics: total fat, saturated fat, polyunsaturated fat, monounsaturated fat, and cholesterol. The mean number of lipid characteristics in ads rises only slightly between 1977 and 1987, but then rises from .13 characteristics in 1987, to .57 characteristics in 1991, before falling to .47 characteristics post-NLEA period.

The mean number of lipids in an ad is the product of two factors, the percentage of ads that have *any* lipid claim, and the average number of lipids in an ad that has at least one claim. The percentage of food ads with a lipid claim grows throughout the period examined here, slowly at first to 10 percent of ads in 1987, then strongly to 34.4 percent in 1991, and further to 39.5 percent of ads in 1997. Thus, the reduction in the mean number of lipids in ads in the post-NLEA period is due to changes in the number of lipid characteristics in ads that have a lipid claim.

The number of lipids in ads with a claim is steady at approximately 1.3 characteristics throughout the late 1970s and early 1980s, rises to 1.65 characteristics in 1991, and then falls back to 1.26 characteristics in 1997. To put this in perspective, in 1977, 1.1 percent of ads have claims for more than one lipid characteristic; by 1983, this has risen to 2.5 percent, by 1991 it rises strongly to 20.1 percent of ads, and by 1997 it has fallen sharply back to 5.0 percent. This evidence highlights the competitive focus on saturated fat and cholesterol claims that rose in the late 1980s before falling back dramatically after the NLEA rules.

- **Competition on Main Nutrients Peaks in 1991** Using a broader index of 12 major nutritional components of foods, we find results similar to those for lipids. The mean number of nutrients featured in advertising begins growing earlier than for lipids, but also peaks in 1991, before dropping 22 percent by 1997.

As with lipids, movement in the overall mean is more the result of changes in the number of different nutrients featured in ads than in the number of advertisements making nutrient claims. If a nutrient claim is made in an ad, the mean number of different nutrients in the ad rises sharply during the 1980s and decreases substantially during the 1990s.

By 1997, the number of nutrients in the average ad with claims has returned to the level of the mid-1980s, a 33 percent drop.

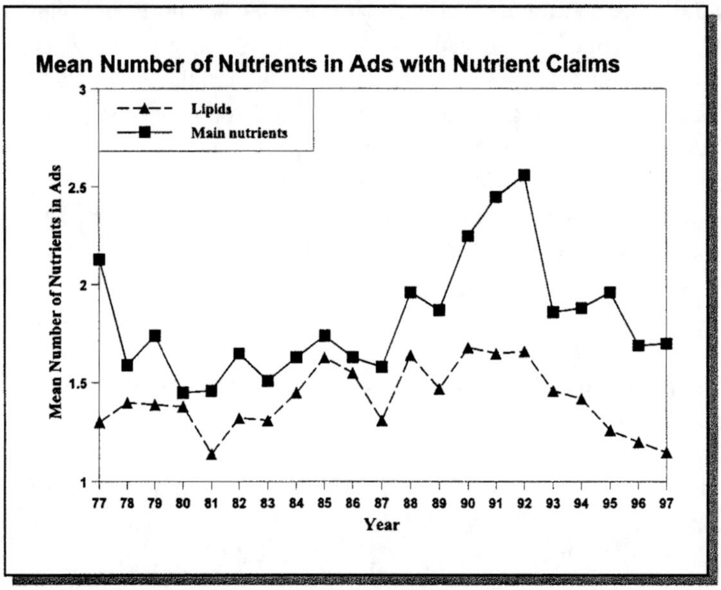

For example, in 1983 4.0 percent of advertisements have claims for 3 or more different nutrients. This rises sharply to 19.9 percent of ads in 1991, before falling back to 8.5 percent of ads in 1997. As with lipids, this evidence suggests that the competitive pressures on nutritional issues of the late 1980s led advertisers to highlight more nutritional characteristics of their products than they had earlier. In the post-NLEA period, nutritional claims in advertising are more limited, focusing on one or two nutrients only. Thus, as for lipids, this evidence provides support for considerable competitive unfolding and does not support the hypothesis that the NLEA environment induces more complete nutrition profiles in ads.

Better understanding of the reasons for these changes and whether they have improved consumer diets are important areas for further research. In earlier work (Ippolito and Mathios, 1996), for instance, we found that the fat characteristics of consumers' diets improved at a faster rate in the late 1980s, compared with the rate between 1977 and 1987. It would be valuable to know whether diets are continuing to improve and at what rate under the policies adopted in the 1990s.

Is There Competition Among "Bads?" The Case of Fats and Oils

The unfolding hypothesis implies that firms with a *relative* advantage over their competitors will be led to advertise that advantage. Thus, even advertisers in "bad food" categories may be induced to focus on nutrition and health as long as consumers are sufficiently aware of nutrition issues and differences on nutrition dimensions *within* the category are sizable.

The Fats & Oils category provides a good opportunity to test the unfolding hypothesis in a "bads" setting. These products generally contain considerable fat, or are substitutes for such products, but they vary substantially in the *type* of fat and in the amount of fat per serving. Heart disease has been linked to some types of fat, particularly saturated fat, cholesterol, and transfatty acids, but not others. So substitutions among fats are important. Under the NLEA rules, direct health claim competition is no longer allowed for products that are not low in fat. Among our findings:

- **Disease and Affiliated Claims Are a Major Feature of Competition in the Fats & Oils Market When Allowed** Even in the late 1970s, between 6.5 and 11.4 percent of fat and oil advertisingper year includes serum cholesterol claims. These claims fade as the FTC Food Rulemaking considers explicit regulation of heart-related claims, but then rise dramatically and immediately to more than 20 percent of ads in 1983, after the end of the rulemaking. By 1988, 45 percent of fat and oil ads include disease or affiliated claims dealing with heart issues, as producers compete aggressively on choices within the category. These claims remain an important feature in the category until 1991, when they fall from 36.7 percent of ads in 1991 to 2.8 percent of ads in 1992, following the November 1991 publication of proposed NLEA rules prohibiting health claims for fat products. After the NLEA rules are in effect, health claims do not reappear through 1997.

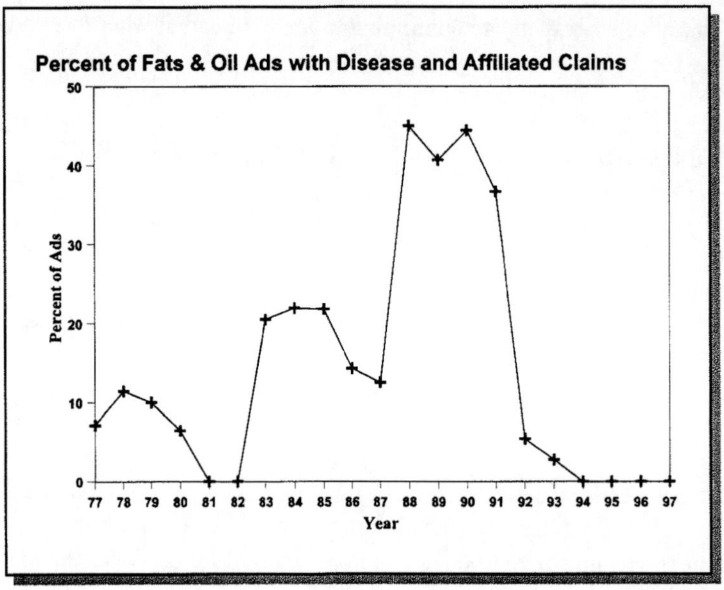

Taken together, this evidence indicates that competition on *bads* can become a major focus of competition in a particular category, as in fats and oils here. Having less of a bad is, of course, a good thing, and apparently advertisers believe that they can communicate the health importance of these differences to consumers in a way that enhances their products' sales.

- ***Advertising For Fats & Oils Falls Dramatically in Post-NLEA Period; Few Compete on Nutrition*** As the focus on health issues fades in the fats and oils category, the amount of advertising also falls dramatically. In 1997, the number of ads is at only 20 percent of its peak level in 1989 and at only 43 percent of its level in 1977. The number of ads with saturated fat claims also drops to near zero. By 1997, few advertisers in the category appear to compete on the health or nutritional characteristics of fat and oil products.

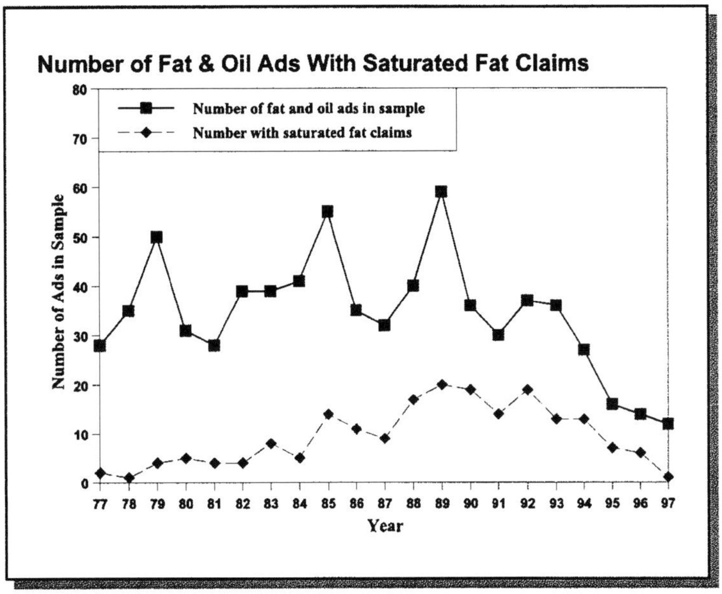

Advertising and Broader Audiences: Do Producers Reach Out with News?

One of advertising's possible strengths is its potential to reach out to consumers with information. As a final test of advertising's information role in markets, we examine advertisers' use of different types of magazines to reach consumers with health news. In particular, we contrast the use of health claims in general readership magazines (*Time*, *Newsweek*, and *Readers' Digest*) with that in women's magazines (*Better Homes & Gardens*, *Good Housekeeping*, *Ladies' Home Journal*, *McCalls*, and *Women's Day*). Women's magazines are the normal magazine medium for food advertising, having 10 times as many ads as the general readership magazines at the start of our period.

- **Food Advertising in General Readership Magazines Increases During Periods of Increased Health Claim Advertising** The idea that producers reach out to the broader audience with health information is supported by data on the number of ads in the two types of magazines. The number of food advertisements in general readership magazines increases following 1987, reaching 140 percent of its 1977 level in 1989, at the height of the health claims period. The number of ads falls in the early 1990s, before rising again in the post-NLEA period. In contrast, the number of ads in the women's magazine sample has been trending downward since the mid 1980s.
- **During Periods of Change, Health Claims Are More Likely in General Readership Magazines** As the regulatory constraints are lifted in the mid-1980s and again after the NLEA rules are in place, the use of disease and affiliated claims rises in women's magazines, but it rises considerably more in general readership magazines, and these magazines have large audiences. In 1989 at the peak, 20.9

percent of all food ads in our general readership magazines contain a disease or affiliated claim compared to 6.6 percent of ads in our women's magazines.

Taken together, these data are generally consistent with the hypothesis that producers will attempt to spread information that expands demand for their products to broader audiences when allowed to do so.

CONCLUDING REMARKS

This report examines a wealth of data on the content of food advertising during the years 1977 to 1997. The data make it clear that nutrition-related claims have become a major feature of food advertising and an important focus of competition. The evidence also makes it clear that regulatory rules and enforcement policy matter – the content of food advertising shifts markedly as the policies towards nutrition and health claims vary over these years.

Among the changes in the post-NLEA period, several findings are worth noting. The nutritional focus in advertising has narrowed substantially. Total fat has become the primary nutritional focus of advertising competition, away from other major nutrients including saturated fat, cholesterol, and sodium. Comparative claims have dropped to very low levels for all nutrients except total fat. For health claims, the most dramatic change has occurred in the market for fats and oils, where competition on the health reasons to choose one fat over another has been eliminated in advertising. The evidence also shows no increased advertising focus on "good foods," and in fact, advertising for fruits and vegetables has fallen significantly since the NLEA.

The ultimate question of which regulatory and legal policies best serve consumer interests requires that we relate the advertising changes observed here to consumers food choices. Until that work is done, this evidence provides us with an important part of that evaluation: objective and detailed information on the content of food advertising under the different policies examined here.

Marketing is often controversial. Producers are trying to sell their products. But marketing claims about important product characteristics – subject to market and enforcement limits on deception – unleash competitive forces that play an important role in shaping the mix of products available in the market and in attracting consumers to products with desired characteristics. As science has shown the importance of nutrition in disease risks, advertising has focused increasingly on nutritional characteristics of food. In crafting policy that serves consumers' interests, it is important that we understand the role of marketing in consumer goods settings. We hope this evidence contributes to that effort.

Chapter 1

INTRODUCTION

Advertising is a prominent feature of consumer good markets. Firms devote substantial resources to advertising as they compete for customers. But advertising remains controversial. Does advertising add to the information base that consumers use to make better decisions or does it lead them to poorer decisions? Does it facilitate competition among firms as they try to meet consumer demands or does it inhibit competition by its costs?

Regulatory and legal rules limit the claims firms can make in their advertising. Primarily the legal standards attempt to prevent deceptive or misleading claims. But what is deceptive and to whom? How do different rules on advertising claims affect firms' incentives to compete? And how do the rules affect the information content of advertising or the dimensions that become the focus of competition in marketing?

In a series of earlier studies,[1] we explore the role of advertising in food markets, especially as it relates to health and nutrition-related claims. Food advertising and labeling has been subject to considerable scrutiny over the last 25 years, and the rules governing food claims have changed several times, making it a fertile venue for study. With the exception of the Pappalardo and Ringold study of the margarine and oil market, our earlier studies have not had direct evidence on the types of claims actually made in advertising. Instead, the studies focus on changes in consumer and firm behavior as regulations change regarding health or other nutrition-related claims.[2]

This study is designed to provide original data on the content of food advertising over a sufficiently long period that both market and regulatory forces can be assessed. More specifically, the study collects a large, systematic sample of magazine food advertising from the leading women's and general readership magazines for the years 1977 through 1997. All advertising for foods in the sampled magazines is included, except that for baby food and alcoholic beverages.

A methodology is developed to extract all nutrition-related claims from these advertisements and to categorize them in ways amenable for study. Some other broad classes of claims, such as taste or new product claims, as well as some health-related pictures and symbols, are also recorded. Each advertisement is coded twice. Computer checking during the coding and computerized reconciliation of discrepancies by a third coder are used to assure

---

[1] See Ippolito, Ippolito and Mathios, and Pappalardo and Ringold (various years) in the references.

[2] Other authors have also examined the impact of labels and advertising, including Teisl, Levy and Derby (1999), Derby and Levy (2000), Weiner (1999), and Moorman (1998).

that the final advertising claim data has a very high degree of accuracy. The data record the presence of claims of various types. The data do not allow us to assess whether the claims are truthful or deceptive to consumers.

After describing the methodology and the sample in more detail in Chapter 2, the study proceeds to provide a wealth of descriptive data on food advertising. Chapter 3 presents data on food advertising spending from public industry sources and demonstrates that the number of advertisements in our sample tracks industry trends quite closely. The chapter also presents data on the amount of food advertising in broad food categories and on trends in nonnutritional types of claims in food advertising.

Chapter 4 presents detailed data on the use of specific nutrient content claims over time for all of the major nutrients, such as total fat, saturated fat, sodium, *etc.*, as well as claims about calories and dieting. The chapter also presents data on the use of general nutrition claims over time, such as the use of the terms *healthy*, *wholesome*, *light*, *etc.*, as well as comparative information on trends in the use of specific versus general nutrition-related claims.

Diet-disease claims and other health effect claims have been particularly contentious features of food advertising. Chapter 5 presents basic descriptive data on the use of these specific health claims, including data on the use of heart disease, cancer, and other disease claims. For the major health claims, the data are also presented by food category.

Regulatory rules for nutrition and health-related claims on food labels and in food advertising have been the focus of considerable debate over the period of our data and have changed several times. Chapter 6 provides a brief review of the major regulatory events related to food marketing, and then uses regression analyses to examine whether and how the use of health and nutrient content claims varies over the various regulatory regimes. The chapter also examines whether the post-NLEA environment is successful in inducing more advertising and a greater health focus in advertising for the foods targeted for increased consumption, such as fruit and vegetables, compared to foods targeted for reduced consumption, such as fats and oils, as had been hoped when the regulations were devised.

Finally, Chapter 7 focuses more directly on some of the economic theories of advertising. The chapter provides evidence on the hypotheses that advertising plays a direct informational role in markets. The chapter also examines the hypothesis that a less restrictive policy towards advertising fosters competitive pressures leading firms to focus on more dimensions of the product in their advertising. The chapter presents evidence on competition on bad as well as good aspects of foods, such as the types of fats, in the fats and oils category. Finally, the chapter examines the hypothesis that firms will attempt to spread information to new audiences through advertising when allowed to do so easily.

This study, like our earlier work, cannot resolve policy questions concerning the use of health information in food advertising and labeling. Instead our goal is to provide a broad range of evidence on the claims advertisers actually make in their food advertising, evidence that has been lacking in previous assessments of the issues. This type of evidence is essential in assessing the types and frequency of various claims flowing to consumers in advertising under different enforcement policies and the reactions of advertisers to the different advertising rules. For instance, if producers do not compete on the nutritional features of their products, concern about stifling that competition would be lessened. Conversely, if certain problematic claims do not occur frequently and regulatory burdens designed to address such claims stifle other informative nutrition-related claims, the regulations might merit reconsideration.

Our primary goal in this report is to provide evidence on advertising content that bears on these and a host of other regulation and policy issues regarding food marketing and consumer policy more generally. In addition, though, the data here should be of interest to marketing researchers, nutritionists, and economists attempting to better understand the forces that shape consumer and firm behavior, especially the role that marketing can play in changing consumer dietary choices.

Chapter 2

METHODOLOGY AND SAMPLE CHARACTERISTICS

INTRODUCTION

The primary goal of this project is to collect reliable data on the types and quantity of nutrition-related claims in food advertising over a sufficient number of years to span recent changes in food labeling and advertising regulations. As detailed in Chapter 3, television is the primary medium used for food advertising. Unfortunately, no archives exist that would allow us to develop a consistent and representative database of food advertising on television. Magazine advertising is the second largest category of food advertising, and many libraries retain copies of the most popular magazines. Since advertising campaign themes are generally carried across the various media used, we would expect changes in advertising claims in magazines to generally reflect overall advertising shifts.

These facts determine the approach adopted here, namely to create a large database of magazine food advertising from the leading magazines used by food producers. A methodology is developed to extract all nutrition-related claims from these advertisements and to categorize them in ways that will be useful for analysis. This chapter describes the sample of advertisements and the methodology for extracting claims from the ads.

ADVERTISING SAMPLE

In developing a sample for this study, we want a collection of advertisements that represents the magazine advertising flowing to consumers and a sample that is consistent over time in order to reveal changes in the content of these advertisements. To achieve these goals, we choose a fixed set of magazines that represents the major sources of magazine food advertising. These magazines are sampled consistently at fixed points during the year for each of the years in the sample. All food advertisements in the selected magazine issues, except for baby food and alcoholic beverages advertisements,[1] are included in the sample.

---

[1] Baby food and alcoholic beverage labels are regulated differently from other foods. Since part of our interest is in describing changes under the NLEA and other food labeling and advertising rules, advertising for these foods is not included. Dietary supplements are also not included in the sample, but diet foods, such as *Slim Fast*, are included. For the remainder of the report, we will use the term food advertising without the qualification that baby food, alcoholic beverages, and supplements are not included.

Years of the Sample

The sample covers the years 1977 through 1997. The sample period begins well after nutrition labeling is put in place in the early 1970s and coincides with the large USDA food consumption survey in 1977. The sample period ends three years after the full implementation of the *Nutrition Labeling and Education Act of 1990* (NLEA) and seven years after passage of the act itself.

These years cover a period of substantial development in the information flowing to consumers from other sources, including the development of dietary guidelines,[2] several changes in food labeling and advertising regulations culminating in the passage of the NLEA, and a number of scientific and public health initiatives involving food issues. These are discussed in more detail as we analyze the data.

Magazines in the Sample

Women's magazines are the primary magazines food producers use for advertising. Circulation data for magazines are available from the Audit Bureau of Circulation.[3] While ranks change somewhat from year to year, the list of the top seven women's magazines is quite stable during the years 1977 to 1997. Using a wide range of libraries in the Washington, D. C. area, as well as purchasing back issues available from the publishers, we attempted to assemble complete samples of the needed issues. This was successful for five of the top six women's magazines, and those five are used in the sample, namely *Better Homes and Gardens*, *Good Housekeeping*, *Ladies Home Journal*, *McCall's*, and *Women's Day*.[4] The same procedure is used to select the top three general readership magazines for the sample. These are *Reader's Digest*, *Time*, and *Newsweek*.[5]

Sampling Frequency

Since most magazines in our sample are issued monthly, we randomly chose a starting month from the first four months of the year and then chose every fourth month for inclusion in the sample. The selected months are February, June, and October of each year. For the weekly magazines (*Time* and *Newsweek*), the second and fourth issues of the month are included. For *Women's Day*, which is issued every three weeks, the first issue of the month is chosen if the selected month has more than one issue. All food advertisements in these issues of the magazines are included in the sample. If an advertisement appears in several magazines, it is included in the sample each time it appears. Thus, the sample measures the quantity of advertising of a particular type flowing to consumers, but not the number of distinct advertising campaigns at any time.

---

[2] The first US Senate *Dietary Guidelines for Americans* is published in 1977; USDA and HHS dietary guidelines begin in 1980 and are revised every five years.

[3] As cited in *World Almanac and Book of Facts*, 1977 through 1997.

[4] The only women's magazine in the top six that is not included in the sample is *Family Circle*. We were unable to find a complete set of the selected monthly issues for this magazine in available libraries. The seventh magazine, *Redbook*, along with other issues of the chosen magazines, are used for testing and development of the coding instrument.

[5] *Modern Maturity* has greater circulation than the selected magazines in the later years of our sample, but this magazine is available only to AARP members.

Summary of Advertising Sample

Table 2-1 summarizes the characteristics of our advertising sample. Our sample includes all covered food advertisements that appear in the February, June, or October issues of the selected magazines for the years 1977 through 1997. This sample has 11,647 advertisements.

METHODOLOGY FOR EXTRACTING AD CLAIMS

Manifest Content Analysis

Claims in the advertisements are extracted using manifest content analysis, which measures explicit advertising claims.[6] In this technique coders are given specific instructions on how to code the words, and to a more limited extent particular symbols or pictures, in the advertisements. Coders are not asked to interpret what the words or pictures might mean in a particular context.

Table 2-1 Summary of Advertising Sample Characteristics

Women's Magazines Sampled *Better Homes and Gardens* *Good Housekeeping* *Ladies' Home Journal* *McCall's* *Women's Day* **General Readership Magazines Sampled** *Reader's Digest* *Time* *Newsweek*
Months and Years Included[1] February, June, October for every year from 1977 to 1997.
Foods Covered All foods except for baby foods and alcoholic beverages.
Advertisements Included All advertisements for covered foods that appear in the selected issues of the selected magazines. Resulting sample has 11,647 advertisements.

Notes. [1] For weekly magazines, the second and fourth issues in the month are included. For *Women's Day*, which is issued every 3 weeks, the first issue of the month is chosen if a month has two issues.

---

[6] Content analysis is a technique used in many disciplines to collect objective, systematic, quantitative, and generalizable descriptions of communication content. The technique has been used to study advertising since at least the 1970s. In recent years researchers have begun to use content analysis to investigate changes in the use of nutrition and health information in food advertising. See, for instance, Lord, Eastlack and Stanton (1987, 1988), Hickman, Gates, and Dowdy (1993), Pratt and Pratt (1995), and Pappalardo and Ringold (2000).

This technique is chosen over latent content analysis, which attempts to measure the effect of claims in the advertisements, because it produces more reliable and objectively verifiable data. Latent content analysis requires those who code the advertising to attempt to interpret what is communicated to the audience, rather than what is actually said in the advertisement. While it is undeniably true that words can have different meanings in different contexts, attempting to discern that meaning in a large scale effort such as this one risks introducing considerable subjectivity and potential bias into the data. The historical nature of the study makes the usual concerns about latent content analysis particularly acute, as coders today would have to infer implied claims consumers would have taken from advertisements decades ago.[7]

Computer-Assisted Coding

After extensive testing with food advertisements outside our sample, detailed instructions were developed to guide coders in extracting claims from advertisements.[8] We use a computer-assisted coding instrument that allows coders to enter data directly into a database. A variety of computerized consistency checks are conducted on each entry as the coding occurs, and coders are not allowed to proceed until any inconsistencies are corrected.[9] Questions are also nested to reduce the burden on coders.[10]

Table 2-2 indicates the major classes of claims coded from the advertising. For each advertisement, a coder is asked if there are any claims of a given type. If the answer is yes, the coder is asked a series of more detailed questions about the claims in the category. For instance, within the category of *General nutrition claims*, coders are asked to indicate the presence of claims in each of the following categories: *Health/Healthy*, *Smart/Right*, *Good/Better for you*, *Nutritious/Nutrients*, *Wholesome*, *Enriched/Fortified*, *Light/Lighter*, *Lean/Leaner*, *Guilt free/No guilt*, *Fresh*, *Natural/No artificial/Real/Pure*, *Energy*, *Young/Fitness/Well-being*, and *Other general nutrition claims*.

---

[7] See Ringold and Calfee (1989) and the papers cited there for a recent discussion of the two techniques.

[8] Coding instructions are available from the authors upon request.

[9] For instance, if a coder indicates that an advertisement has no fat claims, but then tries to code a *low fat* claim, the program stops the coding until the inconsistency is corrected.

[10] For instance, a coder would be asked if there were any fat claims of any type. If the coder answers no, all of the more specific fat claim categories are filled in appropriately and the coder is directed to the next channeling question.

Table 2-2 Major Categories of Coded Claims[1]

General nutrition claims	
(*e.g., nutritious, healthy, natural, light*)	
Nutrient content claims	
(*e.g., fat claims, saturated fat claims, calcium claims, calorie claims*)	
Health claims	
(*e.g., heart disease claims, cancer claims, serum cholesterol claims, bone claims*)	
Expert/Dietary guidance claims	
(*e.g., American Heart Association recommends, dieticians recommend, dietary guidelines given*)	
Auxiliary health information given	
(*e.g., total diet context, need for exercise*)	
Health symbols or pictures	
(*e.g., USDA Food Pyramid, health association seal, people exercising*)	
Other ad claims of the following types:	
Taste, aroma, texture	*Suggestions for use*
New, introducing, improved	*Convenient, quick, easy*
Variety	*Price, cost, coupon*
Promotional offer	

Notes. [1] A more precise definition of each category of claims is given when the results are discussed in later chapters.

Figure 2-1 provides a copy of the coding instrument for the fat claim questions as it appears on the computer screen. To illustrate the coding process, consider this series of questions. Coders are first asked whether the advertisement contains any *(total) fat claims*. If it does not, coders are immediately channeled to the next category of questions. If the ad does contain fat claims, the coder is first asked whether there are any *fat level claims*. If the answer is yes, the coder proceeds to characterize the claim(s) into the categories *No fat/Free,*[11] *Low fat,*[12] *% Fat free,*[13] and *Other level claim.*[14] Coders then record whether the amount of fat is given quantitatively, whether alone or with another claim.

[11] Coders are instructed to code a claim in this category if it is a *no fat* claim, including *no fat, fat free, zero fat, nonfat, zip, nada, none, without any, 100% fat free, 0% fat*.

[12] Coders are instructed to code claims in this category only if the terms *low* or *lowfat* are used.

[13] Coders are instructed to code any *X% fat free* or *X% fat* statements in this category, except for *100% fat free* and *0% fat*, which are coded in the *No fat* category.

[14] Coders code any other statement describing the absolute level of fat in the product in this category. The text of such claims are recorded in the text box.

Figure 2-1 Illustration of Claim Coding Questions

Fat Claim Questions

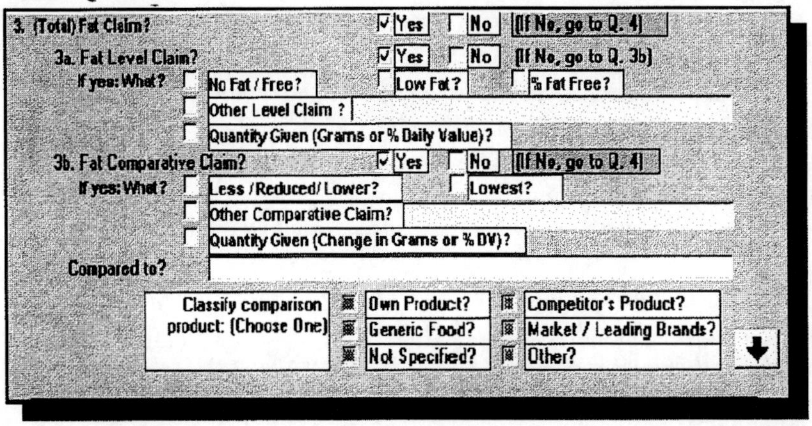

Health Claim Questions[1]

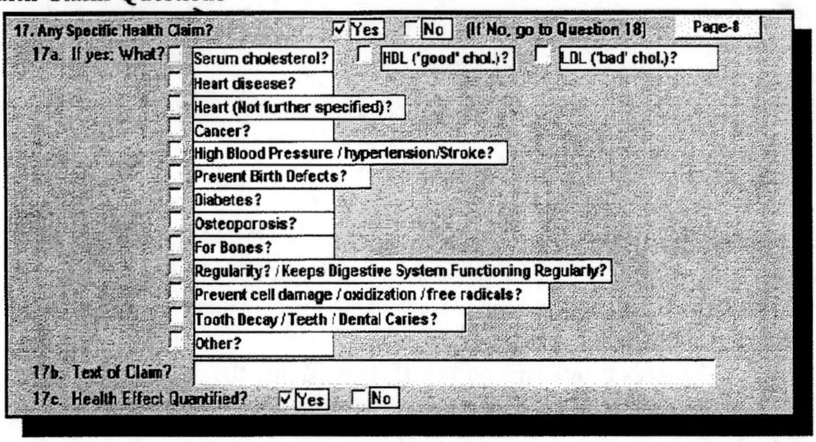

Notes. [1] The complete coding questionnaire is provided in the appendix.

The coder is next asked whether the ad has any *fat comparative claims*. If not, the coder is taken to the next category of claims. If there are comparative claims, the coder is asked to classify the claim(s) in the categories *Less/Reduced/Lower,*[15] *Lowest,*[16] and *Other comparative claim.*[17] Coders then record whether the quantity is given in the comparison (either in grams or % Daily Value) and the comparison product (if any). Finally, the coder is asked to classify the comparison product as *Own product, Competitor's product, Generic food, Market/Leading brands, Not specified,* or *Other.* Claims for other nutrients are coded in the same general way.

---

[15] Claims are coded in this category only if they use the terms *less*, *reduced*, or *lower* (possibly with the quantity given, as in *50% less fat*).

[16] Claims are coded in this category only if they use the terms *lowest* or *least*.

[17] All other claims that compare the amount of fat using comparative terms are coded in this category, such as *compare*, *side-by-side nutrition tables*, *fat has been cut*, *etc.* The text of such claims are recorded in the text box.

Figure 2-1 also includes the section of the coding instrument that captures specific *Health claims*. These claims, discussed in more detail in Chapter 5, are defined as any statement or term in the ad referring to specific health effects of nutrients or foods, such as statements about specific disease risks or any other specific health effects of foods.[18] As shown in the figure, a coder is asked whether the ad contains any specific health claims. If not, the coder is channeled to the next set of questions. If there are health claims, the coder records the claim(s) in the listed categories. The text of the claim is recorded and an indicator is set if the claim somehow quantifies the health effect.

The entire coding instrument is included in the appendix. *General nutrition claims* are coded first, since pretesting indicated that this broad category of claims is the most challenging for coders, and the coding is more accurate if the category is addressed first. *Nutrient content claims* are coded next for each of the major nutrients, as illustrated for fat above. A catchall category is presented for *Other nutrient claims* not covered elsewhere to ensure that all nutrition claims are captured. *Calorie and diet claims* are next, followed by *Health claims*, and other miscellaneous health-related categories, such as *Expert and dietary advice claims*, *Auxiliary health information*, and *Health symbols or pictures*. Finally, broad categories of other types of claims in the ad are coded, such as *Taste, aroma, texture claims* and *New, introducing, improved claims*.

Nine coders participated in the project. All were undergraduate students or college graduates employed by the FTC, and all were facile with computer technology. All received extensive training in the coding rules and in the use of the computerized coding instrument. Each had to reach a level of proficiency before undertaking actual coding.

Black and white copies were made of all advertisements to ensure that the availability of color ads does not affect historical trends.[19] Each advertisement is coded independently by two coders. Coders code all advertisements in assigned magazine issues. These issues are randomized over time and magazine to ensure that learning and maturation effects in the coders do not affect historical trends or magazine category results.[20] A computer matching program compares the two codings and any discrepancies are resolved by a third independent coder. These resolved data are used in the study.[21] The authors did not do any of the coding or resolution.

Coders are instructed to consider all words in the advertisements that are readable, including words on the labels shown in the advertisements. Thus, no distinction is made

---

[18] This category includes specific disease claims, such as "Concerned about heart disease? Brand X is low in saturated fat and cholesterol." The category also includes claims referred to as *structure-function claims* in labeling regulations, such as "to maintain healthy cholesterol levels," "for strong bones," or "Fiber helps keep your digestive system functioning for natural regularity." This category does not include general health or nutrition terms, such as *natural* or *wholesome*.

[19] For many of the older magazines, we had to make copies of the advertising at outside libraries, where color copiers were not available.

[20] We began coding a smaller set of four magazines, the two leading women's magazines (*Better Home and Gardens* and *Good Housekeeping*) and the two leading general readership magazines (*Reader's Digest* and *Time*). Once we were assured that costs were acceptable, we expanded the set of magazines to include those described above.

[21] Prior to resolution, reliability rates between coders are very high, ranging from 98 percent or higher for specific nutrient claims and 99 percent or higher for specific health claims, to 93.8 percent or higher for general nutrition claims and 90.8 percent or higher for the broad classes of nonnutritional claims. The bulk of the disagreements in the initial codings involve one coder missing a claim; only a very small portion of the disagreements involve substantive questions about how best to code the claim under the coding rules. Detailed reliability rates are available from the authors.

between "primary" claims and other claims in the ad. All distinct claims in the ads are recorded and many advertisements have multiple nutrition-related claims.

OTHER INFORMATION ABOUT THE ADVERTISEMENTS

Certain other basic information about each advertisement is also recorded in the database.[22] In particular, the data include the magazine, month, and year of the ad, the company/brand of the product (as best it can be determined from the ad), the name of the food (text), the number of products in the ad, the size of the advertisement, whether the nutrition label is shown in the ad, and food category indicators to categorize the product(s) in the advertisement into basic food categories. Size is recorded in one of four categories: 1/4 page or less, more than 1/4 page to less than one page, one page, and more than one page.

CONCLUDING REMARKS

To our knowledge, this study provides the most comprehensive examination of magazine advertising content ever undertaken. The final sample includes detailed content assessment of 11,647 food advertisements that cover a 21 year period. To put this in perspective, Ringold and Calfee's (1989) cigarette advertising study is based on a sample of 568 ads over a 60 year period, and Pappalardo and Ringold's (2000) study of oil and margarine advertising is based on 412 advertisements over 40 years. The study's scale is possible, in part, because it made use of computer technology to reduce the burden on coders and to increase the accuracy and speed of the data coding and verification process.

We believe the advertising dataset developed for this project presents a very accurate and complete picture of the types of nutrition-related claims made in magazine food advertising over the years 1977-1997 covered by the study.

[22] Since these data do not involve claims, they were recorded by one of the authors (Ippolito).

Chapter 3

BROAD TRENDS IN FOOD ADVERTISING

INTRODUCTION

Before turning to nutrition-related claims, this chapter provides some background information on general trends in food advertising during the years 1977-1997. Data are presented on overall advertising expenditures for foods and on trends in advertising spending by medium. These data from industry sources are compared to trends in our sample of magazine advertising and are found to be quite consistent. The chapter also provides some general information from our sample on advertising frequency over time and for different food groups. Information is also provided on the use of certain broad categories of claims in food advertising, such as taste, product use, or convenience claims. Information on these claims provides some context for assessing changes in the health-related claims that are the focus of the following chapters.

BACKGROUND

Industry Data on Food Ad Spending

Advertising expenditures are available by medium for a variety of industries from Leading National Advertisers (LNA). Table 3-1 gives advertising spending for food products as reported by LNA for each year and gives the percentage of that spending in magazines, television, and other covered media. The *Television* estimates are the sum of network, spot, cable, and syndicated advertising for the available years as reported by LNA. The *Other Media* category is the sum of Sunday magazine supplements, newspaper, outdoor, network radio, and national spot radio. Advertising expenditures for *Food and Food Products (F100)* and for *Confectionery, Snacks & Soft Drinks (F200)* are summed for the table. This includes all food advertising in LNA except for alcoholic beverages, and thus, corresponds quite well to the foods covered in our sample.

Table 3-1 Advertising Expenditures for Food Products by Media

Year	All Media ($1000)	Magazines (%)	TV (%)	Other Media (%)
1977	1,697,723	8.9	88.2	2.9
1978	1,950,534	9.6	88.2	2.2
1979	2,240,247	9.0	88.4	2.7
1980	2,305,795	9.2	88.0	2.8
1981	2,463,145	9.4	87.4	3.2
1982	2,766,022	9.5	87.3	3.2
1983	3,010,569	9.7	87.7	2.6
1984	3,377,074	9.9	87.6	2.5
1985	3,757,174	10.5	85.9	3.7
1986	3,895,879	11.3	85.5	3.3
1987	4,047,054	10.4	85.9	3.8
1988	4,712,518	9.1	80.9	10.0
1989	4,595,836	10.6	84.7	4.7
1990	5,020,701	10.1	82.7	7.3
1991	4,699,701	10.4	83.3	6.3
1992	4,742,045	11.1	82.6	6.3
1993	4,725,794	11.2	82.2	6.6
1994	5,347,362	11.1	83.3	5.7
1995	5,461,176	14.0	80.0	6.0
1996	5,684,047	13.6	80.1	6.3
1997	5,986,885	13.6	79.5	6.9

Data. BAR/LNA Multi-media Service, *Ad $ Summary*, Leading National Advertisers, New York, NY, annual. Data is summed for *Food and Food Products* (F100) and *Confectionery, Snacks & Soft Drinks* (F200). TV includes network, spot, cable, and syndicated. Other media includes network and national spot radio, Sunday magazines, newspapers, and outdoor advertising captured by LNA.

As shown in Table 3-1, spending for food advertising is disproportionately concentrated in television throughout the period. This percentage falls from 88 percent in 1977 to just under 80 percent in 1997. In contrast, the percentage of food advertising spending in magazines, while a much smaller portion of total advertising, increases over this period, from nearly 9 percent in 1977 to 13.6 percent in 1997. Spending on food advertising in all other media combined is 10 percent or less of the total, and usually under 7 percent, during the period. Like magazines, spending in these other media tends to increase over the period, rising from 2.9 percent in 1977 to 6.9 percent in 1997.

Figure 3-1 Cost of Advertising in Magazines and Network TV[1]

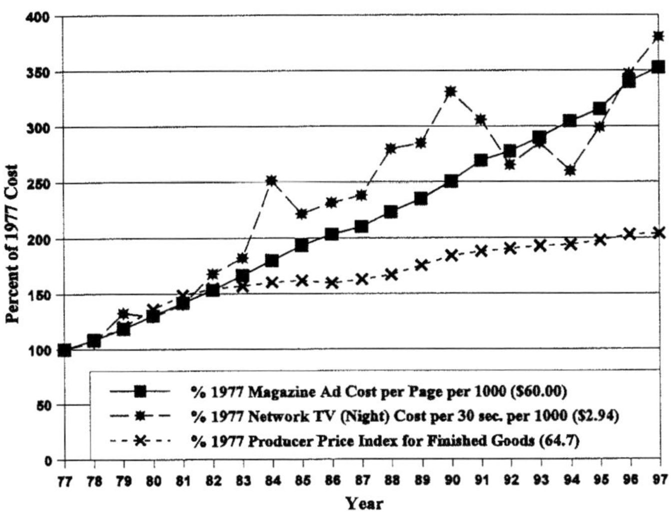

Figure 3-2 Number of Food Ads and Ad Pages in Sample[2]

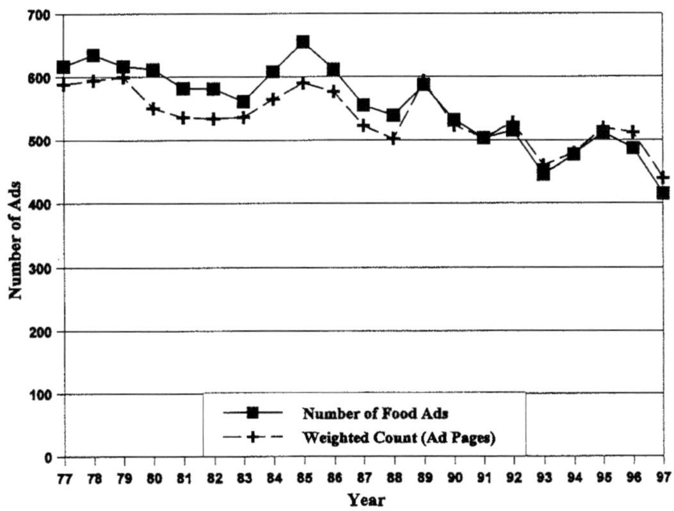

Notes. [1] Network nighttime television cost per 1000 households from Neilsen Media at Television Bureau of Advertising, www.tvb.org/tvfacts/index, 1998, (TVB); magazine ad cost for a 4-color ad per page per 1000 circulation from Magazine Publishers of America, also at TVB, and Producer Price Index for Finished Goods [100=1982] from Bureau of Labor Statistics, in *Economic Report of the President, 1999.* 1977 values in parentheses.

[2] Number of Food Ads is a count of ads in the sample per year. The Weighted Count multiplies each ad by the number of pages in the ad before summing.

The amount of advertising and the media mix for advertising in any product category depends in part on the absolute and relative prices of advertising in the media best suited for reaching the target audience. The costs of advertising in some media increase markedly in real terms over this period. Figure 3-1 illustrates the cost of reaching 1000 households on network television (nighttime, Monday through Saturday) and the cost a 4-color ad per page per 1000 circulation for magazines.[1]

The figure indicates that the price of advertising has increased rapidly over the period in these two media, which are heavily used for food advertising. Moreover, these costs have increased faster than other producer costs since the early 1980s. For instance, the Producer Price Index for Finished Goods[2] is also shown in Figure 3-1 and increases at a much slower rate than either the magazine or TV advertising costs after 1983. Using this index as a price deflator, between 1977 and 1997 the real cost of advertising increases by 73 percent for magazines and by 87 percent for TV.

Comparison With Food Advertising Sample

Figure 3-2 depicts the number of food advertisements in our sample of magazines over time. The number of ads per year falls during the period, which might be expected given the increasing real price of magazine advertising during most of the period. In 1977 our sample contains more than 600 advertisements; by 1997 this number falls to approximately 400 advertisements.

Figure 3-3 Size Distribution of Ads

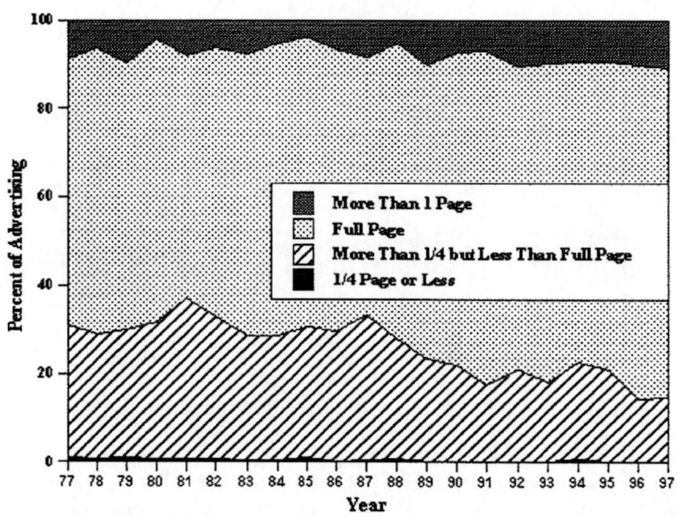

[1] Network television cost per 1000 households is from Nielsen Media Research, February each year, available from Television Bureau of Advertising (TVB), 1998, at www.tvb.org. Magazine advertising cost is for a 4-color ad per page per 1000 circulation from Magazine Publishers of America, also available from TVB.

[2] The *Producer Price Index for Finished Goods* is from the Bureau of Labor Statistics as reported in *Economic Report of the President, 1999.*

The size of food ads in our sample, measured as the total number of pages of space taken up by the ad, also rises somewhat over time, from an average of .95 pages in 1977 to 1.05 pages in 1997. Figure 3-3 gives the size distribution of advertisements over the period. In 1977 approximately 65 percent of ads are one page or more in size, but by 1997 this figure is 85 percent. Most of this shift is due to the reduced use of smaller sized ads after 1987.

Figure 3-4 Magazine Ad Expenditures for Food Advertising: Sample Versus Industry Data

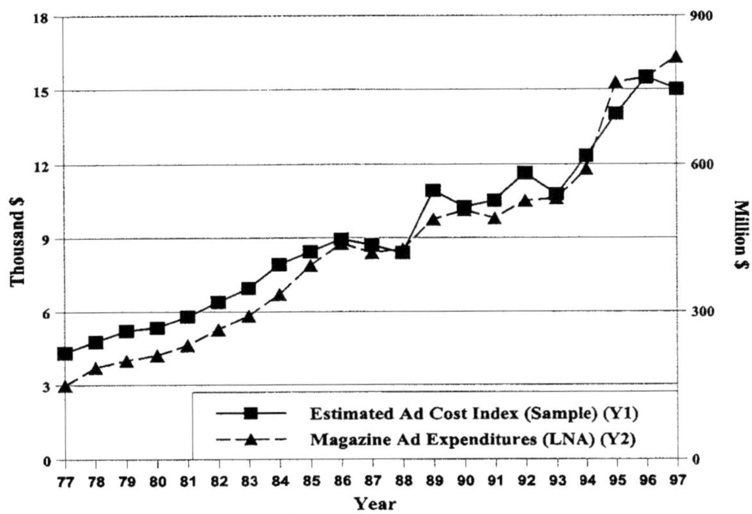

Notes. [1] The estimated ad cost index for the sample is computed as the weighted count of ads per year (page count) multiplied by the cost of a 4-color ad page per 1000 circulation.

[2] Industry data on magazine advertising expenditures taken from Leading National Advertisers data for *Food and Food Products (F100)* and for *Confectionery, Snacks & Soft Drinks (F200).*

To correct for this change in the size of ads, Figure 3-2 also includes data on the estimated number of advertising pages per year in our sample, that is, the size-weighted average of the count data in each year. With either measure, the level of advertising in our sample falls substantially over the period.

As one means of verifying that this sizable downward trend in advertising in our sample is consistent with overall industry expenditure data for food advertising, we create a cost index for our sample by multiplying the number of pages of advertising in each year by the cost of a 4-color ad per 1000 circulation in that year.[3] This index is illustrated in Figure 3-4 together with the LNA estimates of magazine advertising expenditures for foods (the combined F100 and F200 categories described above). As is clear from the figure, the two data series follow a very similar pattern, suggesting that our sample mirrors changes in the overall market data quite well.[4]

[3] Virtually all advertising in our sample is colored.

[4] Since we do not have annual circulation data for the magazines in our sample, we cannot calculate a full cost estimate for the sample. In particular, our index does not reflect any changes in circulation over these years. Magazine circulation data from Magazine Publishers of America for the 50 leading magazines (available at the

Advertising and Food Categories

To provide information on the distribution of advertising across food categories, Table 3-2 lists definitions for 9 major food categories that cover the universe of food products in the sample. Because of their breadth of coverage, these food categories each include a wide variety of foods. The definitions follow the USDA's practice for food categorization, in which mixed foods are assigned to categories based on their primary food ingredient. Thus, for instance, a hamburger made up of meat, bread, and other miscellaneous ingredients is treated as a meat mixture, since meat is the primary ingredient.[5] The *Dressings/Etc.* category is a catchall category that includes small categories of foods not captured elsewhere, such as meal replacement products (*e.g.*, liquid meal products) and general brand advertising (which is a very small part of the sample).

Table 3-2 Description of Food Categories

Food Category	Description
Meat/Eggs/Mixtures	Beef, pork, lamb, veal, game, bacon, sausage, franks, lunch meats, and substitutes; eggs and substitutes; mixed foods with these items as the major ingredient, including sandwiches, stews, meat in sauces, frozen dinners, etc.
Poultry/Fish/Grain/ Mixtures	Chicken, turkey, and other poultry; fish and seafood; rice, pasta, stuffing; mixed foods with these items as the major ingredient, includes Italian, Oriental, Mexican food, etc. Soup.
Cereals/Breads	Ready-to-eat and cooked cereals; breads, rolls, croissants, bagels, English muffins, flour, etc. Does not include bread used in sandwiches, etc., if sold as a mixed food. Also pancakes and waffles.
Dairy	Milk, milk-based drinks, and powdered milk; cheese, cream, sour cream, yogurt and substitutes. Does not include these items in mixed foods, as in lasagna or cheeseburgers; does not include dairy desserts.
Fats & Oils	Butter, margarine, spreads, lard, shortening, oils, etc.
Desserts/Salty Snacks/ Sweet Breads	Ice cream, ice milk, and substitutes; pudding, jello; cakes, cookies, pies, and related baked goods; sweet sauces, jelly, candy, gum, etc.; nuts, seeds, and peanut butter; crackers, salty snacks, chips, popcorn, etc.; sweet breads, muffins, coffee cakes, donuts, nut breads, danish, etc.
Fruits/Vegetables/Juice	Fruit, vegetables, juice, dried fruit, tomato sauce, beans, potatoes, etc.
Soft Drinks/Coffee/Tea/ Other Drinks	Coffee, tea, soft drinks, fruit drinks (not juice), etc. Does not include milk, juice, or alcoholic beverages.
Dressing/Sauces/Gravy/ Seasonings/ Misc.	Salad dressing and mixes, meat flavored sauces, white sauce, gravy, cooking sauce, soy sauce, etc. Seasonings, vinegar, yeast, cornstarch, pectin, unflavored gelatin, baking soda, etc. Meal replacements, sugar and sugar substitutes, general brand advertising.

TVB site in footnote 1) shows overall circulation rising approximately 5 percent between 1977 and 1986, remaining approximately stable until 1993, and then falling about 8 percent in the mid-1990s. Assuming our sample of magazines followed this pattern, these circulation changes are not large enough to alter our conclusion.

[5] The categories here are amalgams of the food categories used in Ippolito and Mathios (1996), designed to link to USDA food categorizations and described in detail there.

Table 3-3 indicates the percentage of food ads in each year that contain a product from the listed categories. The table lists 1977, 1987, and 1997 data, but unless otherwise indicated, is generally reflective of the data over the whole period. The percentages do not add to 100 percent in any given year, because some advertisements contain more than one product, including products from different categories.

As shown in the table, food advertising is spread over the full gamut of foods, but it shows considerable stability over time with a few exceptions. Both the *Fats & Oils* and the *Fruit/Vegetables/Juice* categories show systematic reductions in advertising in the post-NLEA period. The *Desserts/Etc.* category fell somewhat in the mid-1980s and then stabilized. With these exceptions, there is little systematic trending in the annual data for these broad food categories, a finding that is also reflected in the three years of data reported in the table.

Table 3-3 Percentage of Advertising by Food Category

Food Category	1977	1987	1997
Meat/Eggs/Mixtures	10.0	8.3	8.4
Poultry/Fish/Grain/Mixtures	19.1	15.9	20.2
Cereals/Breads	10.0	9.4	10.6
Dairy[1]	4.7	7.6	14.0
Fats & Oils	4.5	5.8	2.9
Desserts/Salty Snacks/Sweet Breads	22.9	20.9	20.5
Fruit/Vegetables/Juice	16.4	18.4	11.8
Soft Drinks/Coffee/Tea Other Drinks	8.6	9.2	9.6
Dressings/Sauces/Gravy/Seasonings/Misc.[2]	15.9	16.6	8.7[3]

Data. Data from advertising sample; product category definitions in Table 3-2. Totals do not add to 100 percent, because some advertisements include products from more than one product category.

Notes. [1] The data for dairy products is quite variable from year to year. The data for these three years do not accurately indicate the trend in the annual data, which is only slightly positive.

[2] Miscellaneous products and general brand advertising are a small part of this residual category, together never rising above 3 percentage points in the total.

[3] The 1997 data for Dressings, etc. is anomalous; in 1996 16.2 percent of the ads included products from this category. The annual data shows only a very slight downward trend over the years 1977-1997.

EVIDENCE ON BROAD CATEGORIES OF AD CLAIMS

Our primary interest in this report is with the many types of claims related to nutrition, diet, and health. However, in coding the advertisements in our sample, we also collected some information on several other classes of claims frequently observed in food advertising. These other categories of claims provide some context in which to judge the frequency of nutrition-related claims over time. They also can act as controls in assessing whether changes observed in nutrition-related claims are specific to those types of claims or simply reflect other forces in advertising affecting the frequency of all types of explicit claims.

Taste, Aroma, Texture Claims

Food advertising frequently contains claims about the sensory aspects of the product. In our study, coders were instructed to record the presence of a sensory claim whenever an ad made "an *explicit* claim about taste, aroma, or texture" of the product.[6] Figure 3-5 illustrates the percentage of ads per year in our sample with a *Taste, aroma, texture claim*. These data indicate that sensory claims are very common in food advertisements: more than 80 percent of all ads contain such claims between 1977 and 1986, and while the percentage falls somewhat in the latter half of the period, it remains above 65 percent of all ads throughout the period.

Convenient, Quick, Easy Claims

Figure 3-5 also presents the percentage of ads that contained a *Convenient, quick, easy claim*. For this category of claims, coders were instructed to record the presence of "any claims about the product as convenient, quick, or easy to use."[7] Approximately 40 percent of all ads in the sample include such claims in 1977, and while the frequency of these claims falls somewhat over time, it remains above 25 percent of all ads throughout the period.

Figure 3-5 ***Taste, Aroma, Texture*** **and** ***Convenience Claims***[1]

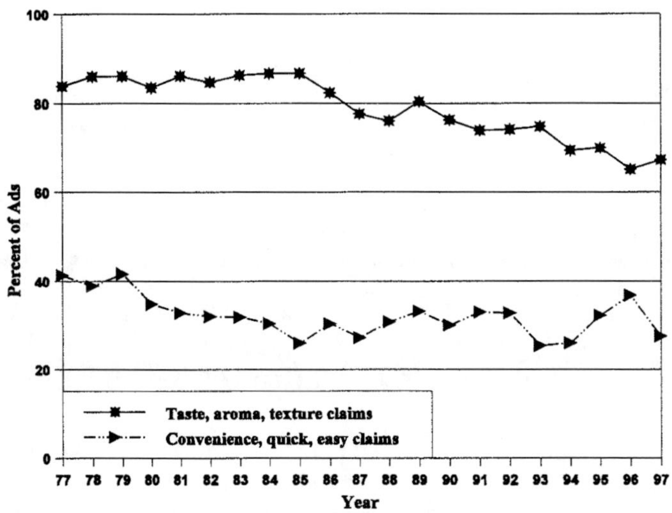

[6] Coder instructions included examples or added guidance for coding sensory claims, including for taste claims: the examples "'tastes good, 'Umm, umm good,' *etc.*" for aroma claims: "Any claim about the smell of the product;" and for texture claims: "Any claim about the texture of the product, such as smoothness, crunchiness, creaminess, richness, *etc.*"

[7] Coders were guided to record any ad that includes a claim that "specifically focuses on the little time to prepare, the ease of preparation, convenience, instant, ready-to-drink, *etc.* ... a 'portability' claim, as in 'goes anywhere' or 'take it with you,' *etc.*" Coders were specifically instructed *not* to judge whether the time listed in recipes is short.

Figure 3-6 ***New, Introducing, Improved*** **and** ***Variety Claims***[2]

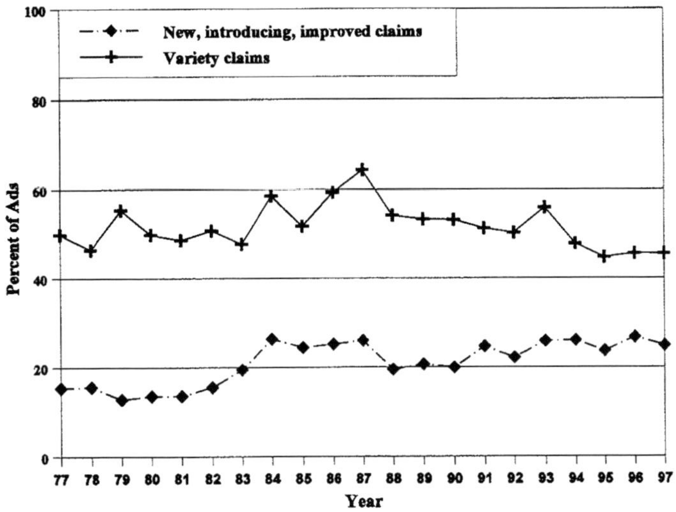

Notes. [1] *Taste, aroma, texture claim* category includes all sensory claims of these types. *Convenience claim* category includes all claims about ease of use. See text for definitions.

[2] *New, introducing, improved claim* category includes all claims about new or improved products or varieties. *Variety claim* category includes all claims about the varieties available, including package sizes, flavors, etc. See text for definitions.

New, Introducing, Improved Claims

Another type of claim that is relatively common in food advertising and that might be related indirectly to new health information is a claim that a product, or version of a product, is new to the market or has been improved in some way. For this category of claim, coders were instructed to record the claim "only if the ad makes any claim about a *new* or *improved product* or *new varieties*." As shown in Figure 3-6, *New, introducing, improved claims* become somewhat more common over time. Approximately 15 percent of all ads in our sample contain a *New, introducing, improved claim* in 1977, and this rises to more than 25 percent by 1984, where it remains in 1997.

Variety Claims

Many food ads contain information about varieties of the product available in the market. This includes claims about different flavors, package sizes, and types (*e.g.*, regular versus instant, regular versus a low calorie version, *etc.*). As shown in Figure 3-6 approximately 50 percent of all ads in our sample have a *Variety claim* in 1977. This rises to over 60 percent by 1987 and falls to 46 percent by 1997.

Suggestions for Use Claims

Many food advertisements contain recipes or other explicit suggestions for using the product. Coders were instructed to code all *explicit* suggestions for using the product whether as a food or for other uses (*e.g.,* baking soda for odors, applesauce instead of oil for baking, *etc.*). Coders were explicitly instructed not to code pictures of the product in use, if there was no explicit text claim indicating a use of the product. As shown in Figure 3-7, approximately 50 percent of all advertisements in 1977 have a *Suggestions for use claim.* The percentage begins falling in the early 1980s, reaching a low point of approximately 30 percent in 1989 and 1990 before rising again somewhat. Throughout out the period, suggestions for use are a common feature of food advertising.

Price, Cost, Coupon Claims

Another relatively common type of claim in food advertising relates to the relative economy of the product. Straightforward price claims are not a typical feature of this type of food advertising, since prices are set at the retail level and these advertisement are typically manufacturer-sponsored advertisements. For our study, coders were instructed to code a *Price, cost, coupon claim* only if "the ad contains a cost-related claim, *e.g., economical* or *only 30 cents per serving, etc.*, or the ad contains a coupon or mail-in rebate or coupon offer." As shown in Figure 3-7, *Price, cost, coupon claims* have trended down over the years of this study. Approximately 20-30 percent of food ads contain such claims in the late 1970s and early 1980s, but by the middle 1990s, less than 10 percent of food ads in our sample contain such claims.

Promotional Offer Claims

Another category of claims that is relatively frequent in this type of food advertising is *Promotional offer claims*. For this category, coders were instructed to record claims "*only if* the ad contains a *contest, mail-in or call-in promotional offer* for goods, recipe books, *etc.*, offers of *any other good* with the product (*e.g.,* "free toy inside"), offers a *special reusable package* (*e.g.,* a historic or decorative tin box), or if *recipes* are offered on or in the package, *etc.*" As shown in Figure 3-8, approximately 25 percent of all food ads in the sample contain *Promotional offer claims* in the late 1970s, but this falls over the 1980s and early 1990s, before rising substantially to 30 percent in 1997.

Figure 3-7 *Suggestions for Use* and *Price, Cost, Coupon Claims*[1]

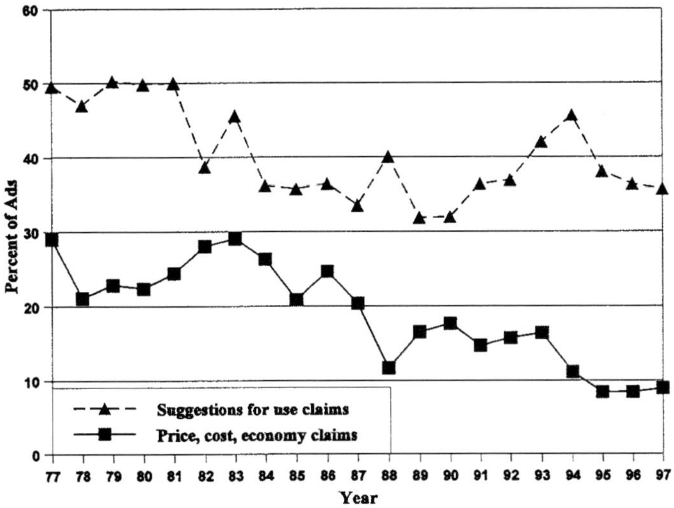

Figure 3-8 *Promotional Offer Claims*[2]

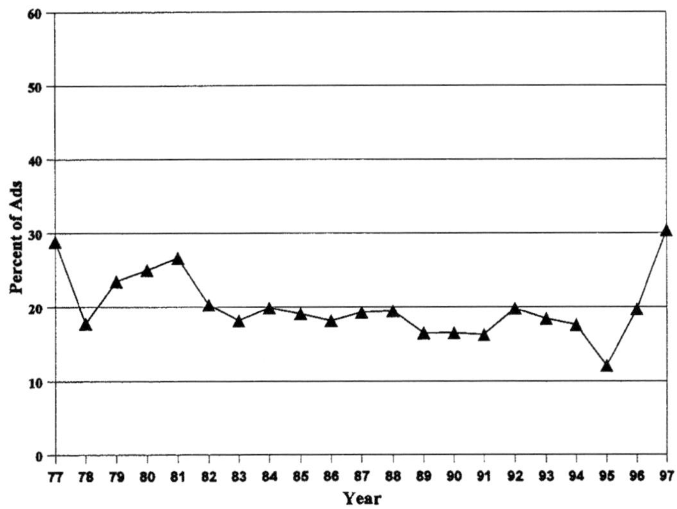

Notes. [1] *Suggestions for use claims* category includes all explicit suggestions for using the product, including recipes. *Price, cost, coupon claim* category includes all cost related claims, such as *economical* or the presence of coupons in the ad. See text for definitions.

[2] *Promotional offer claims* category includes all contests or other promotional offers.

Summary

The data for these categories of claims generally show relatively stable use over time with modest downward trends in most cases. These findings are illustrated by the simple linear trend regressions shown in Table 3-4. All of the estimated trends are negative except for the category of *New, introducing, improved claims*, which has a positive trend. Moreover, all of the trends are statistically significant except that for *Variety* claims.[8] The *New, introducing, improved claim* category is the one category that may relate to changes in the use of nutrition-related claims, because the introduction of products formulated to present a better nutrition profile is often accompanied by such *New claims*.

With this background on food advertising overall, we now turn to the use of nutrition claims in advertising over the years 1977 to 1997.

Table 3-4 Linear Trend Regressions for Broad Classes of Other Claims[1]

Class of Claims [Percent of Ads w/Claim]	Trend	Constant
Taste, aroma, texture [79.6]	-.010 (-16.42)**	.90 (125.12)**
Convenient, quick, easy [32.4]	-.004 (-5.89)**	.37 (43.55)**
New, introducing, improved [20.7]	.006 (9.92)**	.15 (20.04)**
Variety [51.8]	-.001 (-1.42)	.53 (58.82)**
Suggestions for use [40.6]	-.007 (-8.72)**	.47 (53.53)**
Price, cost, coupon [19.5]	-.009 (-15.22)**	.29 (40.70)**
Promotional offer [20.2]	-.003 (-4.43)**	.23 (31.77)**
Any of these claims [96.5]	-.003 (-9.08)**	.99 (304.97)**

Data. Individual ad data from advertising sample. 11,647 advertisements total.

t-statistics in parentheses.

Notes. [1] Linear model of the form D = Constant + a_0 (Time-77), where D=1 if ad includes claim in class, 0 otherwise; Time is a date variable in the form 77.17 for February 1977, ..., 97.83 for October 1997. Results are comparable in probit estimates and in probit models controlling for key regulatory periods as in Chapter 5, except for promotional claims, where the trend in the latter case is positive but not significant.

[8] These results are comparable in probit estimates and in estimates that control for the major regulatory periods, as described in Chapter 5, with the exception that the trend for *Promotional offer claims* is no longer significant in the latter case.

Chapter 4

NUTRITION CLAIMS IN FOOD ADVERTISING

INTRODUCTION

Nutrition-related claims have been a part of food advertising throughout the years 1977 to 1997, but the types, specificity, and frequency of claims have changed in a variety of ways over the period. This section presents basic descriptive data on the use of specific nutrient content claims in advertising for all of the major nutrients and for calories over the years of our sample. We also present detailed information on the use of general nutrition claims, such as *wholesome*, *nutritious*, *healthy*, *etc.*

Evidence on the use of specific disease-related claims, often called *health claims*, is presented in Chapter V. Analysis of whether and how the use of nutrition and health claims varies systematically with regulatory changes is presented in Chapter VI.

NUTRIENT CONTENT CLAIMS

Definition

In our coding scheme, *Nutrient content claims* are defined as "any statement or term in an ad referring to a specific nutritional characteristic of a food, *e.g.*, 'low fat,' 'high fiber,' 'contains vitamin E,' 'low calorie,' *etc.* In this chapter we present detailed data on the prevalence of nutrient content claims for the major nutrients over the years 1977-1997, and where claims are widely used, some breakdown of the specific type of claim made.

Lipid Claims

First we consider the class of *Lipid claims*, which includes total fat, saturated fat, cholesterol, and other fat and oil claims. This class of claims is one of the most frequently observed in advertising and one that has generated considerable controversy.

Total Fat Claims

The category of *Total fat claims* includes all claims about unspecified types of fat, such as "low fat" or "contains only 6 grams of fat." Coders are specifically instructed not to include claims about types of fat in this category, such as claims about saturated fat, polyunsaturated fat, or monounsaturated fat. Other types of specific fat or oil claims, such as "made with canola oil" or "contains no animal fat," are also not included in the total fat category. Each of these is coded separately.

In 1997, total fat claims are by far the most frequently used nutrient content claims in food advertising. As illustrated in Figure 4-1, 36.9 percent of all food ads in the sample contains a total fat claim. Also as illustrated in the figure, this focus on total fat in advertising is a relatively recent phenomenon. Prior to 1987, the percent of advertising that contains a total fat claim never rises above 5.0 percent of ads. After 1987, the prevalence of total fat claims in food advertising grows steadily, slowing temporarily in the early 1990s, but then continuing on the same growth path.

Figure 4-1 Percentage of Ads with *Total Fat Claims*[1]

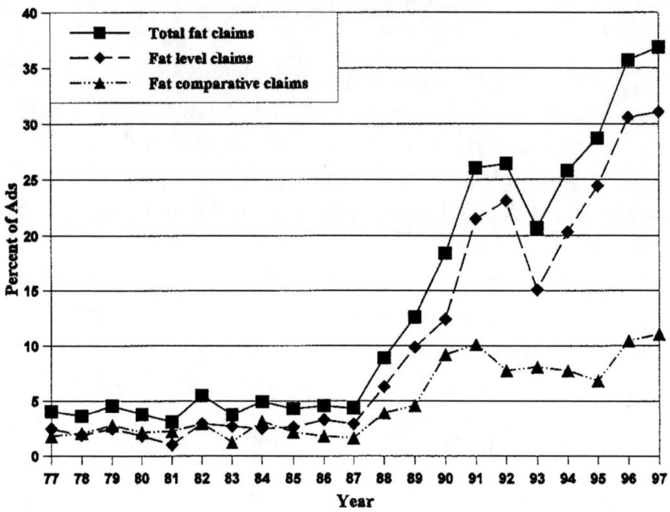

Figure 4-2 Percentage of Ads with *Saturated Fat Claims*[2]

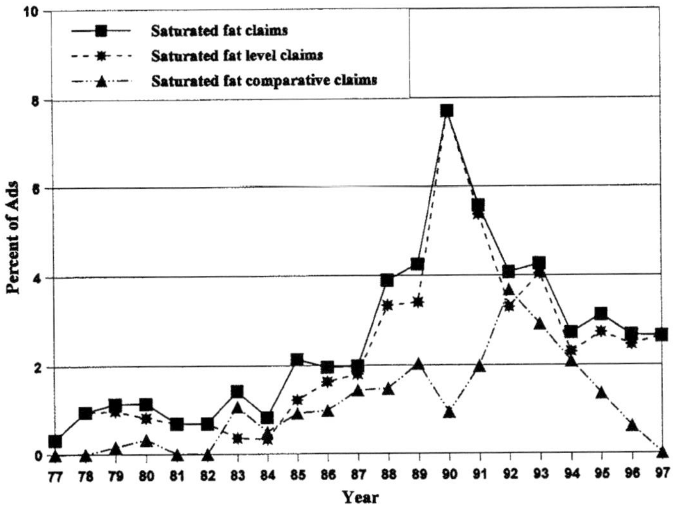

Notes. [1] *Total fat claims* include all claims about unspecified types of fat in the product, such as "low fat." The category does not include saturated fat claims or claims about other specific types of fat or oil in the product. *Fat level claims* are quantitative or qualitative claims about the level of fat in the product and *Fat comparative claims* compare the level to something else.

[2] *Saturated fat claims* include all statements about saturated fat, whether about the level of saturated fat in the product or the amount compared to something else.

Prior to 1987, *Fat comparative claims* are approximately as likely to be used as level claims.[1] Use of both types of claims also grows sharply in the late 1980s, though use of comparative claims grows less rapidly. After 1990, however, comparative claims stop growing, stabilizing at approximately the 10 percent level. This contrasts sharply with fat level claims which continue to grow rapidly in the 1990s, after a temporary reduction in the early 1990s.

Annual data for these classes of total fat claims, along with comparable data for other major nutrients, are given in Appendix B.

Saturated Fat Claims

Saturated fat is the type of fat most clearly identified with the risk of heart disease. Figure 4-2 illustrates the percentage of food ads that include a saturated fat claim of any type in the years 1977 to 1997. The pattern of use for *Saturated fat claims* over time is distinctly different from that for total fat. Saturated fat claims are used in only 2 percent of advertisements prior to 1987, but then rise in frequency reaching a peak of 7.7 percent of all food ads in 1990, before falling again to under 3 percent of ads in the mid 1990s. Claims about the level of saturated fat in the product follow the overall saturated fat claim pattern quite closely, peaking in 1990 before falling to under 3 percent of ads in 1997. *Saturated fat*

[1] *Fat comparative claims* explicitly or implicitly compare fat in the product to that in some other product, as in "now with less fat." *Fat level claims* refer only to the product itself, as in "contains 6 grams of fat." See Figure 2-1or the coding form in the appendix for the claims in each of these categories.

comparative claims peak at 3.7 percent of advertising in 1992, before falling steadily to zero percent of ads by 1997. Thus, in the post-1990 period saturated fat claims are used much less frequently than in 1990, with comparative saturated fat claims having essentially been eliminated.

Cholesterol Claims

Cholesterol is another lipid component of foods that has been associated with the risk of heart disease. As shown in Figure 4-3, *Cholesterol content claims*[2] vary greatly over the years of our sample. In 1977, only 1.6 percent of advertisements contain any mention of cholesterol content, but by 1987 this percentage rises to 6.7 percent of all advertisements. Between 1987 and 1991, the use of cholesterol claims increases rapidly to 24.7 percent of all food ads in 1991, before falling precipitously to 5.8 percent of ads in 1997. The pattern of increase and then decrease is consistent for cholesterol level and comparative claims, with cholesterol comparative claims essentially having been eliminated by 1994.

Other Types of Fat Claims

Claims for the other major fat components, monounsaturated and polyunsaturated fats, also appear in advertising, but these are relatively infrequent. Claims about the amount of monounsaturated fat in a product never occur in more than 0.9 percent of food advertising in any year of the sample. Similarly, claims about the polyunsaturated fat content of products never exceed 1.8 percent of ads in any year. These claims are more common in advertising for fat and oil products, where the type of fat is a focus of competition.

Figure 4-3 Percentage of Ads with *Cholesterol Content Claims*[1]

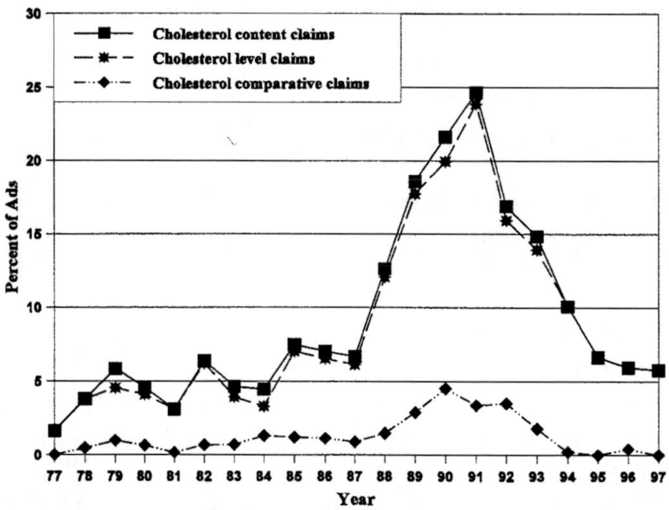

[2] In coding this category of *Cholesterol content claims*, coders are instructed not to include cholesterol statements that are "*clearly* about serum or blood cholesterol." These are coded separately and are discussed in the next chapter.

Figure 4-4 Percentage of Ads with *Other Fat or Oil Claims* and *Corn Oil* and *Canola Oil Claims*[2]

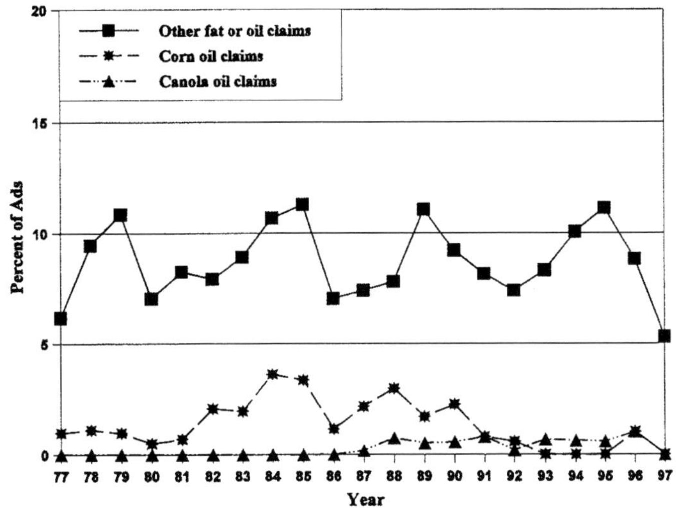

Notes. [1] *Cholesterol content claims* include all statements about the amount of cholesterol in the product, including both level and comparative claims. This category does not include statements that clearly refer to serum cholesterol. These are recorded separately.

[2] The *Other fat or oil claims* category includes any oil or other fat characteristic claim not covered in previous categories. In particular, the category includes type of oil claims, such as corn or canola oil claims, and miscellaneous fat claims, such as "no animal fats," "not hydrogenated," "no transfatty acids," olestra claims, "skim" claims, *etc.*

Advertisers also convey information about the types of fat in products with a variety of other claims. Coders are instructed to record claims if "the ad specifies that the advertised product is or contains some specific type of oil or other fat characteristic that is not covered above." These *Other fat or oil claims* include claims about the type of oil used (*e.g.*, corn oil, olive oil, canola oil), its source (*e.g.*, no animal fats, nondairy, no tropical oils), as well as other characteristics or claims, such as "skim" or "part-skim," "unhydrogenated," "no need to fry," or "baked, not fried."

Figure 4-4 illustrates the use of these *Other fat or oil claims,* as well as the percent of ads that include the specific *Corn oil* and *Canola oil claims*, which are the two most common of the oil claims.[3] The use of these *Other fat or oil claims* does not show any pattern over the period overall, and fluctuates between 5 and 12 percent of all ads.

Summary of Lipid Claims

Overall, the evidence indicates that by 1997 total fat has become the primary focus of lipid claim advertising. Figure 4-5 provides a perspective on the relative incidence of fat, saturated fat, and cholesterol content claims, together with data on the percentage of ads that have at least one claim from *any* of the lipid categories (namely, from the *Total fat, Saturated*

[3] *Olive oil claims* also became more common in the last few years of the sample, appearing in 2.7 percent of all ads in 1997.

fat, Polyunsaturated fat, Monounsaturated fat, Cholesterol, or *Other fats or oils claims* categories). As illustrated in Figure 4-5, the focus of advertising claims for lipids in the 1990s shifts almost exclusively to total fat claims and away from claims about saturated fat, cholesterol, or other types of fat.

Other Major Nutrient Claims

Beyond lipids, a number of other nutrients have been the focus of advertising claims. These include sodium, fiber, calcium, and vitamins and minerals of all types. Interest in these nutrients is in part driven by evolving scientific understanding of the roles they play in generating good health and in reducing disease risks.

Figure 4-5 Percentage of Ads with *Any Lipid Claim* By Major Type: Fat, Saturated Fat, Cholesterol[1]

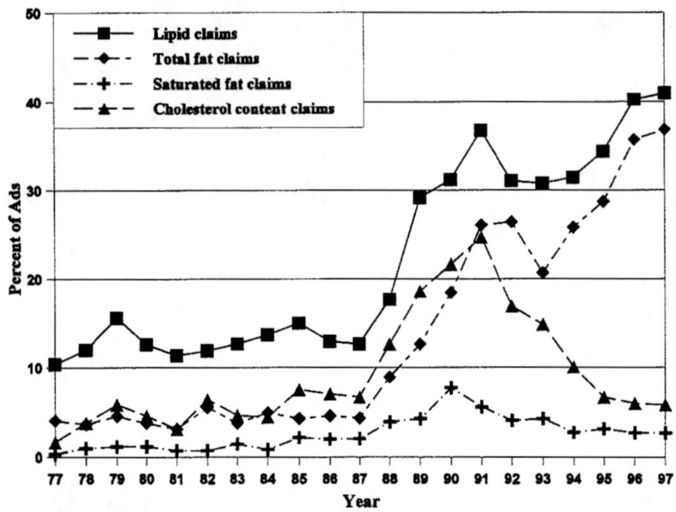

Figure 4-6 Percentage of Ads with *Sodium Claim*[2]

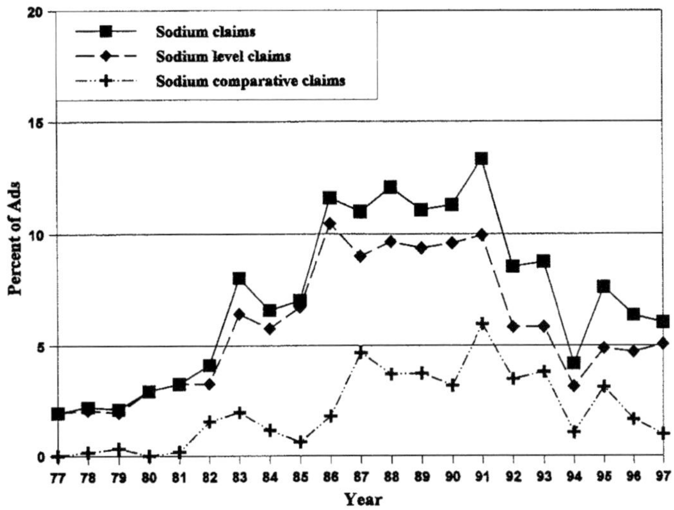

Notes. [1] *Lipid claims* include all fat, saturated fat, monounsaturated fat, polyunsaturated fat, cholesterol, and other fat and oil claims.
[2] *Sodium claims* include all claims about sodium or salt.

Sodium Claims

Sodium is a mineral that has been linked to the risk of hypertension. As shown in Figure 4-6, the use of *Sodium claims* begins to increase in 1980, rising from 2 percent of ads in the late 1970s to 11.6 percent of ads in 1986.[4] The use of sodium claims stays at approximately this level until 1991 when it begins falling. By 1997, the use of sodium claims had been halved to 6.0 percent of food ads. Both absolute and comparative claims fall, but comparative claims fall to very low levels, under 1 percent of ads.

Fiber or Bran Claims

Fiber consumption may be associated with reduced cancer and heart disease risks. Fiber can be soluble or insoluble and occurs in significant amounts in several types of cereal brans, and thus, in whole grains. Coders are instructed to record a *Fiber claim* if the ad contains "any claim about fiber or bran." Coders are specifically instructed to code *whole grain claims* in the fiber claim category.

Fiber claims are most frequent in the bread and ready-to-eat cereal categories, where product names often constitute claims. For instance, the names of the cereals *Fiber One* and *Raisin Bran* are coded as absolute level claims in our data, because the names explicitly refer to the fiber and bran contained in the cereal.[5] As a result, virtually all ads for such products automatically contain a level claim. This is apparent in the data for fiber claims presented in

[4] The Food and Drug Administration began a "sodium initiative" in fall 1981 which leads to the inclusion of sodium as a mandatory element of nutrition labeling in July 1986 (Heimbach 1986).

[5] Recall that *level* claims are defined as claims that describe quantitatively or qualitatively the amount or presence of the nutrient in the product (fiber or bran here).

Figure 4-7. There is virtually no difference between the percent of advertisements containing a fiber claim of any type and the percent containing a level claim. Overall, the percent of ads that contain a fiber claim of some type has been trending up over time, from 2.6 percent in 1977 to 6.5 percent in 1997.

In contrast, comparative claims follow the pattern we have seen for most other nutrients. Fiber comparative claims rise in the mid 1980s, peak at 4.1 percent of all ads in 1989, before falling to near zero by 1993. Fiber claims tend to be concentrated in a few categories, where the pattern is generally consistent with the overall results, but the level of fiber claim use is higher.[6]

Calcium Claims

Calcium has been linked to bone health for many years, but since the mid-1980s, calcium's role in reducing the risk of osteoporosis has been given greater scientific support. Figure 4-8 illustrates the use of *Calcium claims* in advertising. Overall, calcium claims occur infrequently until the mid-1980s when they rise, and with the exception of 1995, stay at approximately 4 percent of all ads through 1997. Like fiber claims, calcium claims are concentrated in a few product categories, such as the dairy category, where they occur in up to 30 percent of ads in some years. Calcium comparative claims, while never very common, have been virtually eliminated in 1997.

Figure 4-7 Percentage of Ads with *Fiber or Bran Claims*[1]

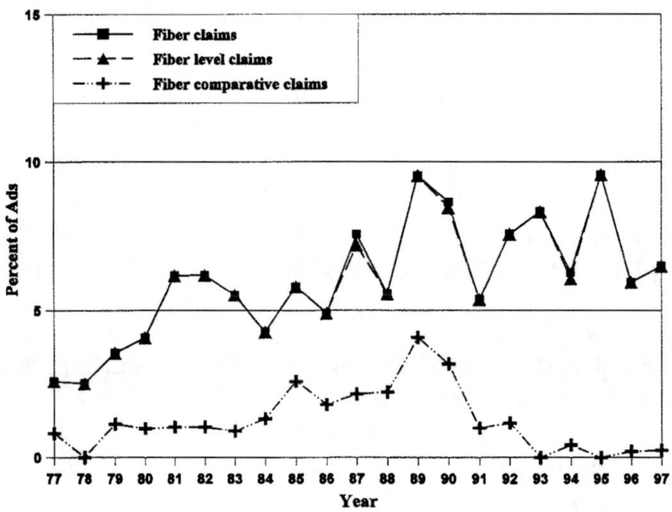

[6] For instance, the data for comparative claims for cereals and breads looks very similar to that for the market overall except that comparative claims peak at 30 percent of all bread and cereal ads in 1990 before being eliminated in the post-1990 period. For absolute claims, the percent of bread and cereal ads with a fiber claim does not continue to increase in the 1990s, but instead falls from a peak of 78 percent of claims in 1989 to 43 percent in 1997.

Figure 4-8 Percentage of Ads with *Calcium Claims*[2]

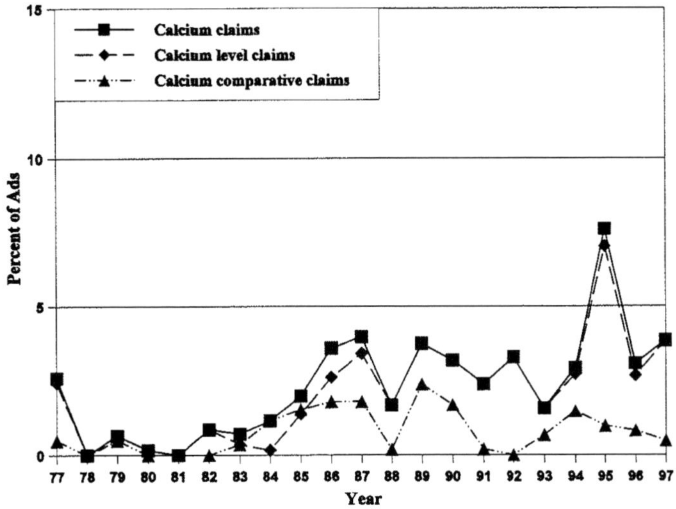

Notes. [1] *Fiber claims* include all claims about fiber or bran. Recall that many cereals and breads have fiber or bran in their names, and thus, almost always have a level claim under our definition.
[2] The *Calcium claims* category includes all claims about calcium.

Vitamin and Mineral Claims

Vitamin and mineral content has been a part of nutrition labeling since its authorization in 1972. In recent years, scientific interest in the role of certain vitamins and minerals in preventing diseases has grown, such as the potential role of antioxidants in heart disease and cancer.

Vitamin and mineral claims include all claims about vitamins or minerals except for calcium and sodium, which are coded separately as discussed above. The category includes general vitamin claims, such as "7 vitamins added," as well as specific claims, such as, "Contains 100% RDA for vitamin C."

As shown in Figure 4-9, the use of vitamin and mineral claims in advertising has fluctuated between approximately 5 and 10 percent of ads over the sample period. Comparative claims have always been less frequent than level claims, but as with most other nutrients, comparative claims have fallen to trivial levels by 1997. The data indicate somewhat greater emphasis on the antioxidant vitamins and on folic acid in later years, but these changes are small. In the mid-1990s, vitamin E claims and folic acid claims are made in approximately 1.5 percent of ads, a rise from near zero levels throughout the 1980s.

Other Specific Nutrient Claims

The data also contain summary information on other nutrients (carbohydrates and protein) or other food characteristics that are related to potential health concerns (sugar and artificial sweeteners, caffeine, and preservatives), as well as a residual *Other nutrient claims* category designed to collect any nutrient claims not captured by our coding scheme. These are summarized below.

Figure 4-9 Percentage of Ads with *Vitamin/Mineral Claims*[1]

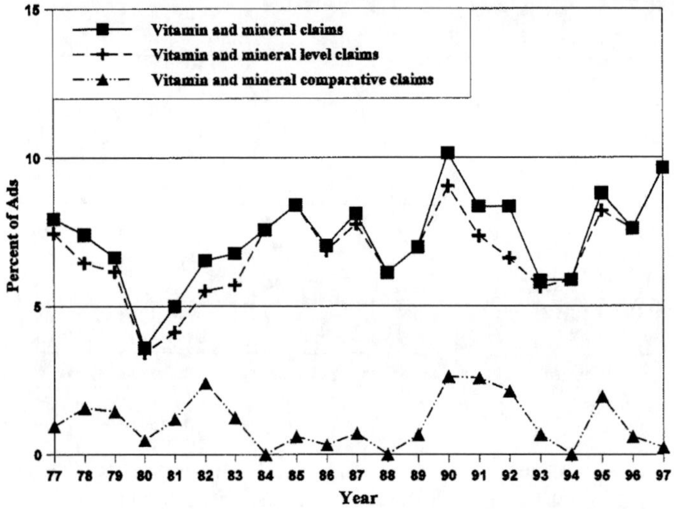

Figure 4-10 Percentage of Ads with *Carbohydrate* or *Protein Claims*[2]

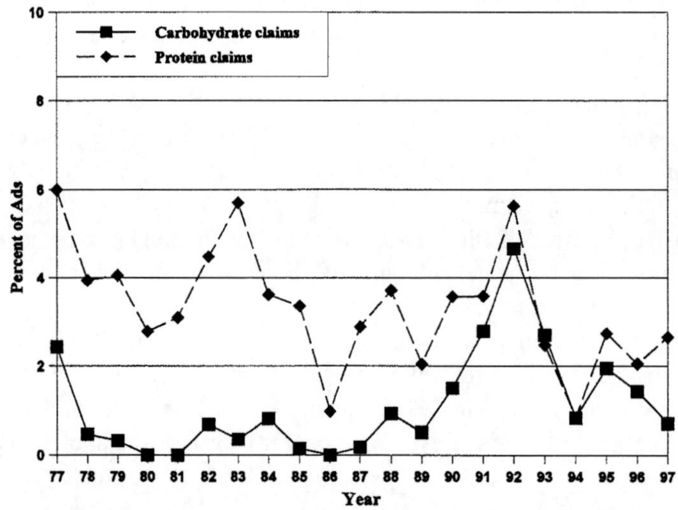

Notes. [1] *Vitamin and mineral claims* include all claims about vitamins or minerals except for calcium and sodium, which are coded separately. The category includes general vitamin claims, such as *7 vitamins added*, as well as specific claims, such as, *Contains 100% RDA for vitamin C.*

[2] Includes any claim about carbohydrates or protein.

Carbohydrate and Protein Claims

Coders are instructed to record any claims about the carbohydrate or protein content of foods. These data are illustrated in Figure 4-10. *Carbohydrates claims* are a feature of food advertising for a few years in the early 1990s, when at their peak, 4.7 percent of ads have a carbohydrate claim. *Protein claims* are a feature of a small percentage of ads throughout the period, with 6.0 percent of ads in 1977 and 2.7 percent in 1997 having a protein claim.

Sugar and Artificial Sweetener Claims

Coders are asked to record any claims about the sugar content of products in the ads, as in "no added sugar," "sugar-free," *etc.* Claims about the use of artificial sweeteners are also recorded, including any references to *NutraSweet* or *Saccharin.* These data are presented in Figure 4-11. The use of both *Sugar claims* and *Artificial sweetener claims* increases in the early 1980s as new artificial sweeteners enter the market, but both fall systematically after 1984. By 1997, 6.3 percent of ads have some type of sugar claim and 1.2 percent of ads have an artificial sweetener claim.

Caffeine and Preservative Claims

Two other small categories of claims are caffeine and preservative claims. Any claims about preservatives are recorded. For caffeine claims, coders are asked to record any claims "about caffeine, including no caffeine, decaffeinated, *etc.*" These data are presented in Figure 4-12. *Caffeine claims* grow from 0.3 percent of ads in 1977 to a peak of 5.6 percent in 1987 before falling over the next ten years to 1.7 percent of ads. *Preservative claims* are more common during the period, but never rise above 6.9 percent of ads. By 1997, these claims have also fallen substantially to 1.9 percent of ads.

Figure 4-11 Percentage of Ads with *Sugar* or *Artificial Sweetener Claims*[1]

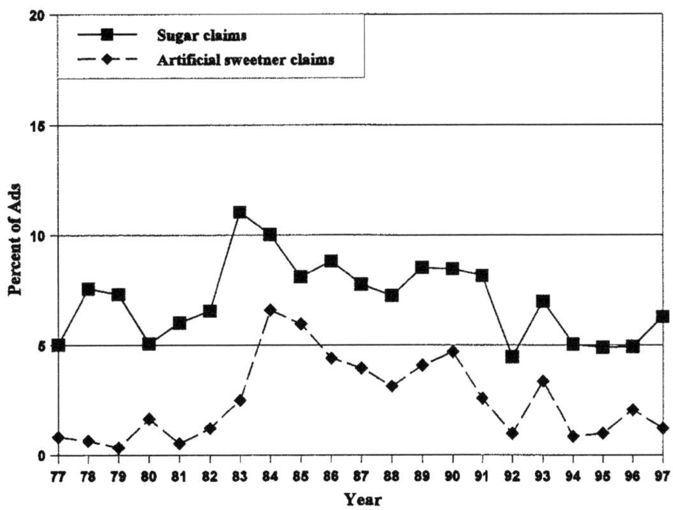

Figure 4-12 Percentage of Ads with *Caffeine* or *Preservative Claims*[1]

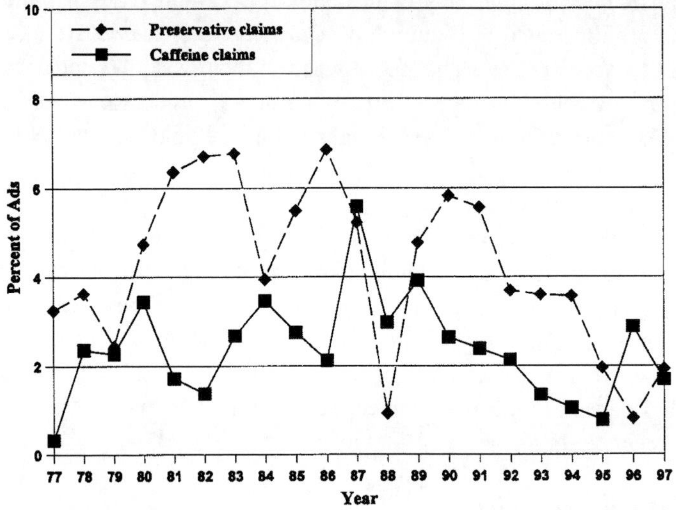

Notes. [1] Includes any claim about sugar or artificial sweeteners, as described in text.
[2] Includes any claim about preservatives or caffeine.

Any Other Specific Nutrient Claims

Finally, coders are asked to record the presence of any other claims about "some other nutrient not covered in any of the above categories or a specific claim about nutrients more generally, (*e.g.*, "All the nutrients of milk" or "7 essential nutrients added," *etc.*)." Detailed examination of the claims captured in this *Other nutrient claim* category indicates that the claims are primarily "no MSG," "wheat germ," "lactose," and "reduced acid" claims, together with general nutrient density or other nutrient comparison claims. Figure 4-13 presents this data and indicates that our coding scheme was quite effective in capturing the vast majority of the nutrition-related claims. With the exception of 1995 when it reached 4.5 percent of ads, the percentage of advertising with any *Other nutrient claims* never exceeds 3.1 percent of advertisements.

Calorie, Dieting, and Weight Claims

Calorie, dieting, and weight claims have long been part of food advertising. To capture these claims, coders are instructed to record a claim if the ad "mentions calories, concern about weight, weight reduction, dieting, *etc.*" In particular, note that foods labeled as *diet* food are coded as diet claims, including, for example, *Diet Coke* or diet margarine.

As shown in Figure 4-14, the percentage of ads with a calorie, diet or weight claim grows steadily from 1977 through 1991 from 7.8 percent of ads to 22.5 percent of ads. Use of calorie, dieting, and weight claims falls in the post-1990 period reaching 12.0 percent of ads in 1997. *Calorie claims* are relatively evenly split between absolute level claims and comparative claims through 1990, when the use of comparative calorie claims falls more rapidly than calorie level claims.

Figure 4-13 Percentage of Ads with *Other Nutrient Claims*[1]

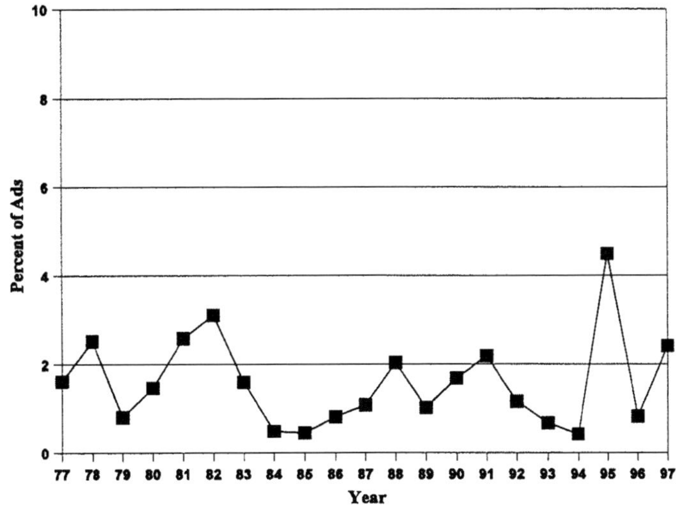

Figure 4-14 Percentage of Ads with *Calorie, Dieting, and Weight Claims*[2]

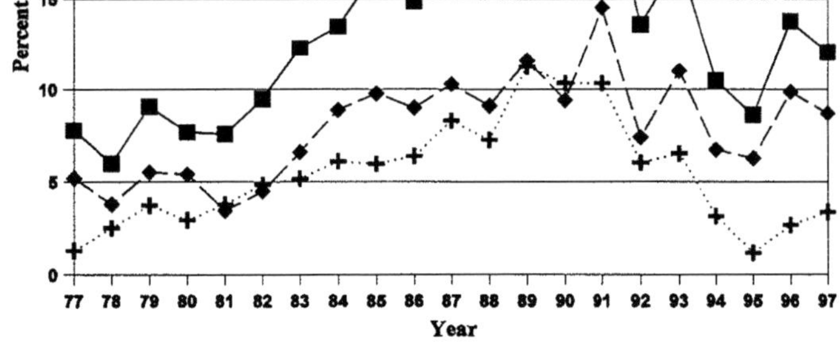

Notes. [1] Includes any nutrient claims not covered elsewhere, as described in text.

[2] *Calorie claims* include all claims about caloric content. *Dieting and weight claims* include all claims about "concern about weight, weight reduction, dieting, *etc.*" Note that the term *diet* would be coded here, including *Diet Coke* or diet margarine.

Figure 4-15 illustrates the percentage of ads that contain a *Dieting or weight claim*. From 1977 to 1991, approximately 5 percent of food ads contained an explicit reference to diet or weight control; after 1991 these claims fall to about 2.5 percent of ads. Thus, the growth and then reduction in the overall *Calorie, dieting, or weight claims* category is attributable primarily to the rise in the pre-1990 period, and then the reduction in the post-1990 period, in the use of calorie claims in advertising.

Figure 4-15 Percentage of Ads with *Dieting and Weight Claims*[1]

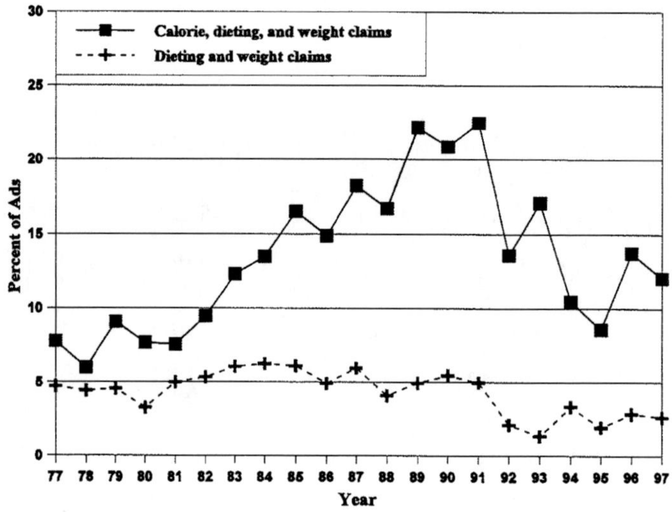

Summary for Specific Nutrient Content Claims

Taken together, several patterns are seen in the use of nutrient content claims in advertising over this period. First, these claims are used throughout the period, but for most nutrients, content claims increase over time in the pre-1990 years, and then decrease in use after 1990.[7] Total fat claims are the notable exception to this finding; total fat claims continue to increase throughout the 1990s. The use of comparative claims also increases prior to 1990 for most nutrients, but these trends are reversed in the 1990s. With the exception of total fat, comparative claims fall to very low levels by 1997 for all major nutrients. We more systematically examine the timing of these changes relative to regulatory changes in Chapter 6.

[7] Very similar patterns are seen in the use of nutrient content claims for new products, as shown in Weimer (1999), Table 1.

Figure 4-16 Percentage of Ads with *General Nutrition Claims* and *Core General Nutrition Claims*[1]

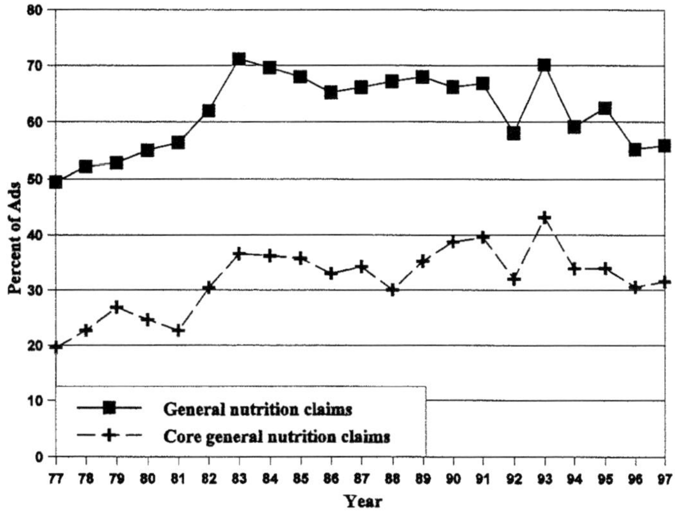

Notes. [1] *Calorie claims* include all claims about caloric content. *Dieting and weight claims* include all claims about "concern about weight, weight reduction, dieting, etc." Note that the term *diet* would be coded here, including *Diet Coke* or diet margarine.

[2] *The Core general nutrition claims* category includes all *general nutrition claims* from the following subcategories: *health/healthy, smart/right choice, good/better for you, nutritious/nutrients, wholesome, enriched/fortified, light/lighter, lean/leaner, youth/fitness/well-being* claims. See text for definitions.

GENERAL NUTRITION CLAIMS

Introduction and Definitions

In addition to claims about specific nutrients and specific nutrition-related characteristics of foods, advertisers use a wide range of more general nutrition claims in their food advertising. Terms such as *healthy* or *wholesome* presumably are used to convey some general nutritional desirability of the food or to invite consumers to check the nutritional profile of the food. But whatever their meaning, this type of claim has been a feature of food advertising throughout the years of our study. In this section we present basic data on the key categories of general nutrition claims.

For our purposes, *General nutrition claims* are defined as any *express* statement or term, other than a nutrient content claim or a health claim (as defined here), that indicates a potential health or nutrition benefit of an advertised food. Coders are specifically asked to code the following subcategories of claims in the general nutrition claims category: *health/healthy, smart/right choice, good/better for you* (in a health context), *nutritious/nutrients, wholesome, enriched/fortified, light/lighter, lean/leaner, guilt free/no guilt/cheating, fresh, natural/no artificial/real/pure, energy claims, youth/fitness/well-being claims*, and any other general nutrition term or statement.[8] Some advertisements contain several general nutrition claims, possibly with other nutrient content claims, and the presence of each type of claim is coded in its relevant subcategory.

Overall Trends

Figure 4-16 depicts the percentage of food advertisements in the sample that contain at least one general nutrition claim. In 1977, 49.4 percent of all the ads have such a claim, and this figure rises to 71.1 percent by 1983, where it stays approximately until declining in the post-1990 period to the 55.9 percent level in 1997. Thus, throughout the period, these general nutrition claims occur commonly in food advertising, and their use increases prior to 1983, and declines in the post-1990 period.

This category of all general nutrition claims includes a very wide variety of claims, including some claims which are more clearly related to nutrition than others. In an effort to explore this broad category of claims, we create a second narrower category of claims, which we will label *Core general nutrition claims*, which includes all general nutrition claims except the *fresh* and *natural/no artificial/real/pure* claims (which might sometimes convey quality more than nutrition to consumers), the *energy* and *guilt free/no guilt/cheating* claims[9] (which may convey calorie or other dieting claims), and the *other* category claims, which is a small category with a variety of claims. Thus, the *Core general nutrition claims* category includes all claims from the subcategories: *health/healthy, smart/right choice, good/better for you, nutritious/nutrients, wholesome, enriched/fortified, light/lighter, lean/leaner,* and *youth/ fitness/well-being* claims

As illustrated in Figure 4-16, the omitted classes of general nutrition claims are a large part of the overall category, but the remaining claims follow approximately the same pattern of change, rising before 1983, then falling slightly in the post-1990 period. After the initial rise, more than 30 percent of all ads had at least one claim from this core category of general nutrition claims.

Natural, No Artificial, Real, or Pure Claims and Fresh Claims

Figure 4-17 illustrates the two categories of claims that are the primary source of difference between the entire class of general nutrition claims and the core class of claims. *Natural, no artificial, real, pure claims*[10] occur in 27.2 percent of all ads in 1977, rise in the early 1980s to 47.2 percent of ads, before beginning a steady decline back to 26.7 percent of all ads in 1997. Similarly, *Fresh claims* rise from 12.6 percent of all ads in 1977 to a peak of 21.2 percent of all ads in 1982 before falling steadily to 8.4 percent by 1997.[11] *Natural, no artificial, real, pure claims* and *Fresh claims* are both used quite frequently in food advertising over this period.

---

[8] Detailed coding instructions are included in the appendix.

[9] *Guilt claims* are coded "if the ad contains any references to removing the *guilt* from eating, or feels like *cheating*, as in *you'll think you're cheating.*"

[10] Coders are instructed to code a claim in this category "if the ad contains any reference to the food or its ingredients as being *natural* or from *nature*, as in *the way nature intended*, any claim that the product does not contain *artificial ingredients* or *chemicals*, or any use of the terms *pure, real, genuine,* or *organic.*" Coders are instructed not to code 100% claims in this category, *e.g.*, 100% beef, 100% cheese.

[11] Coders are instructed to code a claim in the *Fresh* category if there is any claim in the ad "about *fresh* or *freshness*, as in *made with fresh tomatoes.*" The Food and Drug Administration initiated a widely publicized enforcement effort against producers using the term *fresh* in labeling in the early 1990s.

Figure 4-17 Percentage of Ads with *Natural, No Artificial, Real, Pure Claims* and *Fresh Claims*[1]

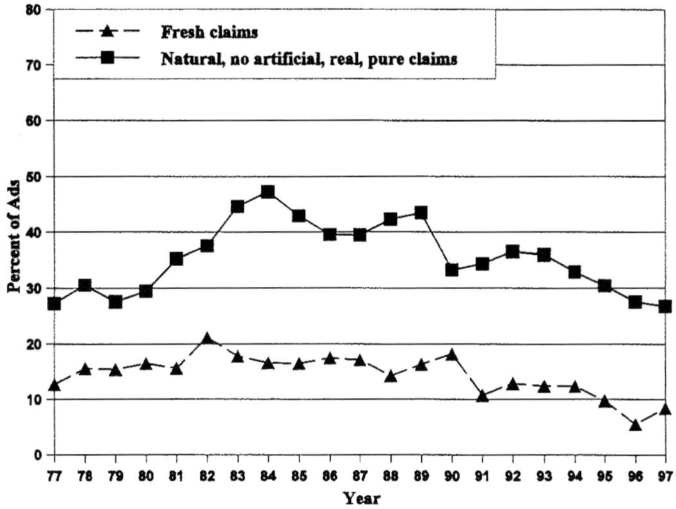

Light and Lean Claims

Within the core category of general nutrition claims are several types of claims that are explicitly regulated under current food labeling rules. As shown in Figure 4-18, *Light, lighter claims* become more frequent in the late 1970s, rising rapidly from 6.3 percent of ads to 15.2 percent of ads but then stay at approximately that level until the 1990s, when their use becomes more variable before falling to 10.1 percent of all ads by 1997. Thus, the evidence suggests that the use of *light* claims may have fallen since 1990, but these claims had been used at a relatively stable level throughout the 1980s.

The use of *Lean, leaner claims* never rises above 4.1 percent of ads in any year during the 1977 to 1997 period, and the evidence indicates no increase in the 1980s.

Figure 4-18 Percentage of Ads with *Light* or *Lean Claims*[2]

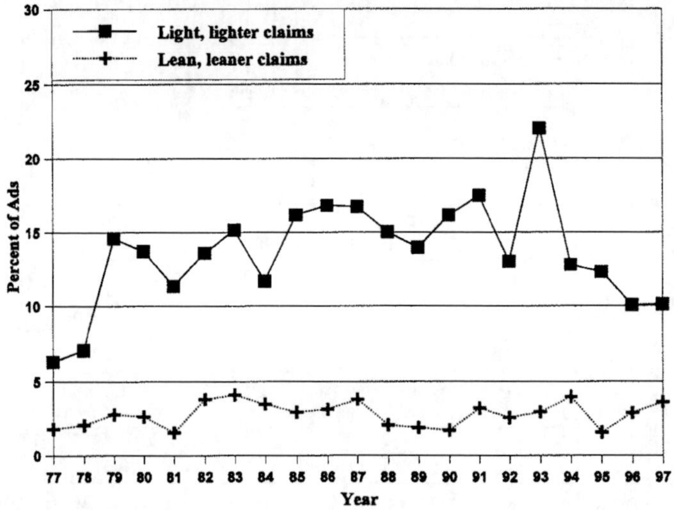

Notes. [1] The *Natural, no artificial, real, pure claims* category includes all natural and related claims, claims about the absence of artificial ingredients, and use of the terms *pure, real, genuine,* and *organic.* See text for further definitions.

[2] Claims using the terms *light* or *lighter* and *lean* or *leaner.*

Health, Healthy Claims

Another type of general nutrition claim that is standardized under current labeling rules is the term *healthy* and related claims, such as "for your health."[12] As shown in Figure 4-19, the percentage of ads using a *Health, healthy claim* rises from 1.6 percent of ads in 1977 to 4.6 percent of ads in 1988. It then rises rapidly to a peak of 12.6 percent of ads in 1990. The evidence also indicates that the use of the term falls somewhat in the post-1990 period, to 8.9 percent of ads in 1997 after the use of the term is restricted under FDA labeling rules. In part, the rise in the late 1980s reflects the introduction of the *Healthy Choice* brand in 1988, but the claim is also used by other advertisers in the late 1980s and early 1990s, sometimes in conjunction with *specific health claims*, as discussed in Chapter 5.

Similar terms, such as *good for you* and *smart* or *right choice*, are also widely used in food advertising to convey "healthy" foods.[13] As shown in Figure 4-19, taken together with *healthy* claims, this group of claims is a relatively significant feature of advertising throughout this twenty-one year period. In 1977, 8.3 percent of food ads contain such claims,

[12] For the *Health, healthy claims* category, coders are instructed to code "if the ad makes any reference to *health, healthy, healthier, healthful, etc.*"

[13] For the *Smart, right choice claims* category, coders are instructed to code "if the ad contains any references to *smart, intelligent, wise, wisdom, right choice,* or related terms when in a health or nutrition context, as in *the smart choice, Smartbeat,* or *you're trying to eat right.*" Coders were instructed to code *Good, better for you claims* "if the ad contains any use of the terms *good, better,* or *best for you* in a health or nutrition context." This is one of the few cases where coders are asked to make a judgment of context, in this case of *good, etc.*, referring to health or nutrition. *Good* and similar terms are often clearly used in other ways, as in "tastes good."

and this figure rises over the years, peaking at 23.3 percent of all ads in 1991. The use of *Healthy, smart, good for you claims* also falls substantially in the post-1990 period to 12.5 percent in 1997.

Figure 4-19 Percentage of Ads with *Healthy Claims* or *Smart, Healthy, or Good for You Claims*[1]

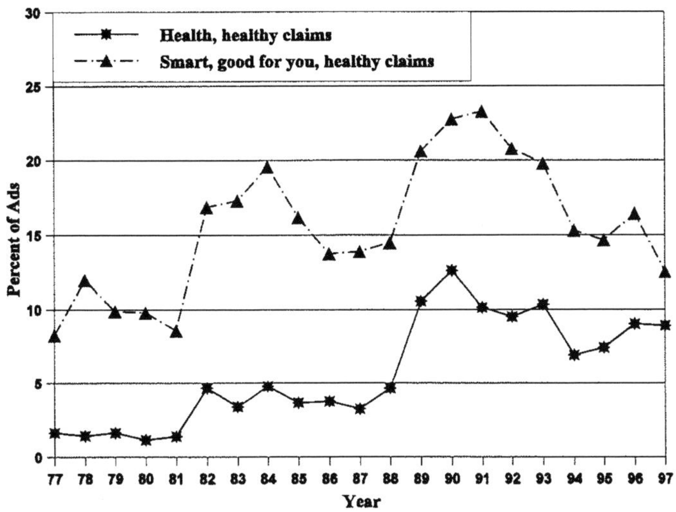

Figure 4-20 Percentage Ads with *Nutritious* or *Wholesome Claims*[1]

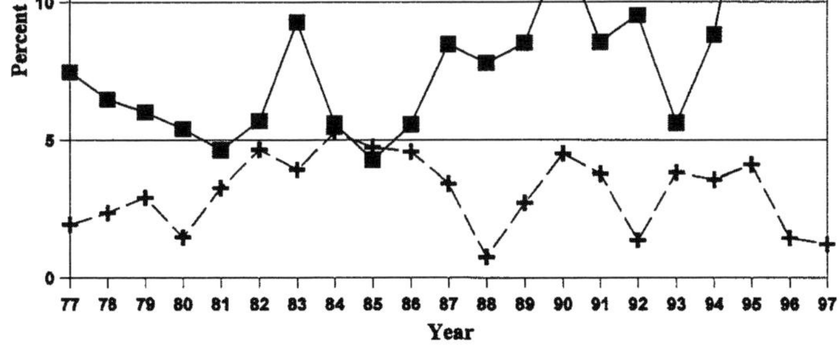

Notes. [1] *Health, healthy claims* category includes claims referring to *health, healthier, healthful, etc.* The *Smart, good for you, healthy claims* category includes the *Healthy claims* as well as claims referring to *smart, intelligent*, and similar terms, and *good, better, best for you claims* in a nutrition context.

[2] The *Nutritious, nutrient claims* category includes all claims with the terms *nutritious, nutrients*, and related terms. The *Wholesome* category refers to claims with that term.

Nutritious, Wholesome, and Enriched, Fortified Claims

Other categories of general nutrition claims that are coded in our data are the broad claims that a product is *nutritious*, *wholesome*, or *enriched* or *fortified* in some general sense.[14] Figure 4-20 illustrates the percentage of advertisements that include claims in the first two categories. Approximately 7.5 percent of all ads contain *Nutritious claims* in 1977, but this percentage falls somewhat until 1981 to 4.6 percent, where it stays approximately until beginning to rise in the late 1980s and continuing through the remainder of our period. In 1997, approximately 10.8 percent of all ads contain a *Nutritious claim*. Use of *Wholesome claims* never rise above 5 percent of all ads and after rising a bit in the early part of the period never shows much significant or systematic movement.

Figure 4-21 illustrates the use of *Enriched, fortified claims*. These claims are required on labeling for certain products with added vitamins or minerals, and some of these label claims are visible in advertising. The percentage of advertising containing *enriched* or *fortified* claims never rises above 7 percent and is typically below 5 percent.

Figure 4-21 Percentage of Ads with *Enriched, Fortified Claims* and *Combined Nutritious, Wholesome, Enriched, Fortified Claims*[1]

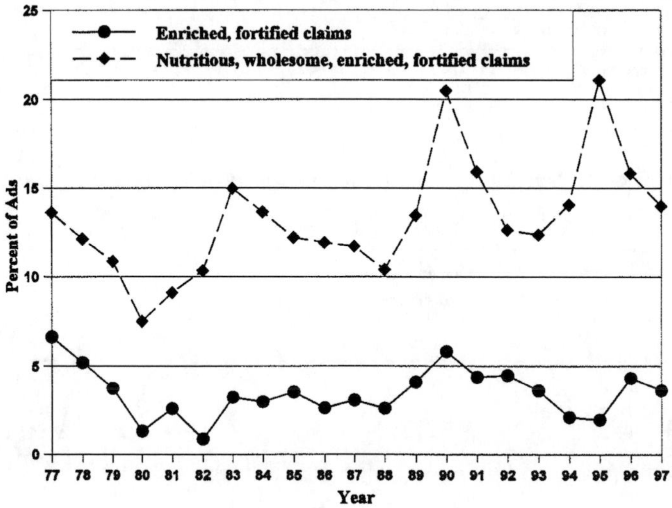

[14] Coders were instructed to code an ad claim in the *Nutritious claims* category if it contained "any of the terms *nutritious, nutrition, nutrients, nourish, nourishment, etc.*" in the *Wholesome claims* category if it contained that term; and in the *Enriched, fortified claims* category if it contained either of those terms, including their use on labels if readable in the ad.

Figure 4-22 Percentage of Ads with *Specific Health or Nutrient Claims* Versus *General Nutrition Claims*[2]

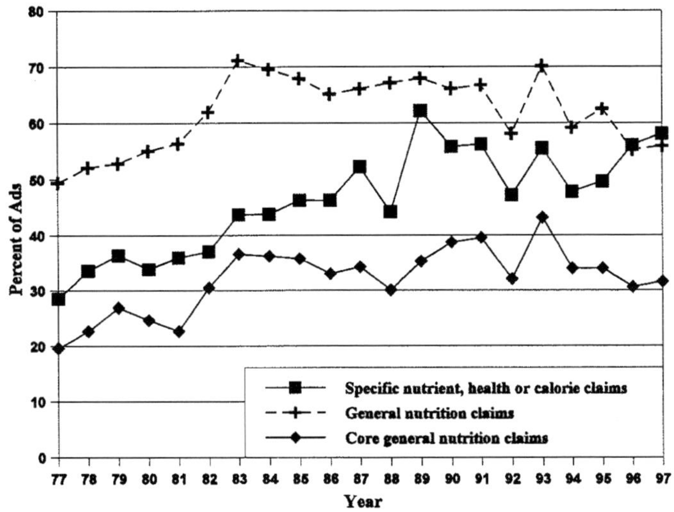

Notes. [1] The *Enriched, fortified claims* category includes all claims with either of those terms. The *Nutritious, wholesome, enriched, fortified claims* category combines the three categories.
[2] See text for definitions.

Figure 4-21 also illustrates the combined use of *Nutritious*, *Wholesome*, and *Enriched, fortified claims*. Combined, these claims show a modest increase over time, with considerable year-to-year variation.

Summary for General Nutrition Claims

Taken together, the evidence on general nutrition claims indicates that these claims are common in advertising. More than 50 percent of ads have a general nutrition claim in 1977, and this percentage rises to more than 70 percent by 1983, where it remains approximately until declining some to about 55 percent in the late 1990s. Claims in the *Natural, no artificial, real, pure claims* category are used most frequently, but claims from the *Fresh*, *Light*, *Nutritious*, and *Healthy* categories are also quite common. If we restrict our attention to the *Core general nutrition claims* category (which includes the *healthy*, *smart*, *good for you*, *nutritious*, *wholesome*, *enriched*, *light*, *lean*, and *youth* subcategories) approximately 20 percent of ads have such claims in 1977, rising to approximately one-third of ads in the early 1980s, where it remains in 1997.

Relative Growth of Specific Nutrition-Related Claims Marks 1977-1997 Years

Finally, in an effort to assess the relative use of specific nutrition-related claims compared to general nutrition claims, we present data that indicates the percentage of ads in each year that contains any *specific* nutrition-related claim.[15] As shown in Figure 4-22, the percentage of food advertisements that have a specific nutrition-related claim shows a sustained and substantial growth between 1977 and 1997. Approximately 28.5 percent of ads have a specific nutrition-related claim in 1977, but this grows steadily to a peak of 62.2 percent of ads in 1989, before essentially stabilizing somewhat below this level in the post-1990 period.

This evidence contrasts with the pattern found in the use of general nutrition claims, which rises before 1983 but then is stable through the 1980s before declining somewhat in the 1990s. Under our broad definition, more ads have general nutrition claims than specific claims for most of the period, but the gap narrows substantially by 1989 and is eliminated by 1997. Under the narrower definition in our *Core general nutrition claims* category, more advertisements have specific nutritionrelated claims in all years between 1977 and 1997. The size of this gap grows in the 1980s, as the use of specific nutrition-related claims increases more rapidly than the use of *Core general nutrition claims*, so that by 1997, advertisements are nearly twice as likely to have a specific nutrition-related claim as a general claim from *Core* category.

Thus, in the broadest sense, the data on the use of nutrition-related claims during the years from 1977 to 1997 indicate a sustained movement towards greater use of specific nutrition-related claims in place of, or in addition to, the use of general nutrition claims that dominated nutrition-related claims in advertising in 1977. As seen in the detailed data, these specific claims grew for most major nutrients in the 1980s but then turned increasingly to total fat claims away from other nutrient claims in the post-1990 period.

[15] In extracting claims from the advertising, coders are led through a series of channeling questions to extract *specific* claims related to nutrition and health issues. The first of these questions instructs the coders to record the presence of "*any* statements or terms related to nutrients, health, calories or dieting" that would be recorded in the specific claims portion of the coding instrument. Thus, the data from this question indicate the presence of any *specific nutrient content* or *health claim*, any *calorie* or *dieting* claim, or any specific fat or oil claim in the ads. This data provides a convenient index of the use of specific nutrition-related claims.

Chapter 5

HEALTH CLAIMS IN FOOD ADVERTISING

INTRODUCTION

Diet-disease claims and other health effect claims have been a particularly contentious feature of food advertising and labeling. These health claims have been a part of advertising throughout the years 1977 to 1997, but their frequency and type have varied both over time and across product groups. In this chapter we present detailed data on the use of health claims during the years of our sample, including data about specific diet-disease claims. Analysis of the relationship between changes in the regulatory environment and the use of health claims is provided in Chapter 6.

DEFINITIONS AND CODING INSTRUCTIONS

For coding purposes a *health claim* is defined as any statement or term in an ad referring to *specific health effects* of nutrients or foods. Thus, this category includes any statements about a disease risk or any other specific health effects of foods. Coders were specifically instructed to code any claims dealing with the following topics: serum cholesterol, heart disease, heart, cancer, high blood pressure/ hypertension/stroke, birth defects, diabetes, osteoporosis, bones, keeps digestive system functioning/regularity, prevents cell damage/ oxidization/free radicals, tooth decay/cavities, and "any other specific disease claim or health-effect claim, as in *to help blood carry oxygen*, *for healthy skin*, or *for gum health*."[1]

This definition is designed to be broad enough to capture all health effect claims, even if they do not actually mention a particular disease. For instance, "helps reduce serum cholesterol" is not actually a disease claim, since heart disease is not actually mentioned, but the claim might be read as a heart disease claim by many consumers who understand that a high serum cholesterol level is a risk factor for heart disease. Similarly, a claim "build strong bones" describes a health effect and not a disease, but might be interpreted to relate to osteoporosis.

---

[1] Detailed coding instructions are included in the appendix.

With these definitions in mind, we now turn to a detailed description of the types and frequency of health claims made during the years 1977 to 1997. Annual data on the use of health claims are provided in Appendix B.

EVIDENCE ON HEALTH CLAIMS

Frequency of Health Claims in Advertising

The top graph in Figure 5-1 depicts the percentage of food advertisements in each year that contain some type of health claim. The percentage of ads that include a health claim remains under 4 percent of advertising until 1987, and then rises to a peak of 11.1 percent of ads in 1989. In the early 1990s the use of health claims falls back to approximately 2 percent of advertisements, rising again after 1994 to reach 8.2 percent of ads in 1997.

To explore the makeup of this class of claims, we categorized health claims into three subcategories: *Disease claims*, which are defined as health claims that specifically mention a disease; *Affiliated claims*, which are health claims that refer to conditions closely affiliated with a disease; and *Nondisease health claims*, which do not meet either of these definitions. *Disease claims* specifically refer to diseases and include claims that mention heart disease, cancer, osteoporosis, birth defects, diabetes, and similar diseases. *Affiliated claims* include claims about serum cholesterol (*e.g.*, "Concerned about your cholesterol?"), heart claims that are not specific as to heart disease (*e.g.*, "heart smart" or "for your heart"), and high blood pressure claims (*e.g.*, "can help reduce high blood pressure.") While not explicitly mentioning a disease, these claims are all closely affiliated with particular diseases. Finally, the *Nondisease health claims* category includes a variety of other health effect claims, including bone claims that do not mention osteoporosis (*e.g.*, "build strong bones"), claims about regularity, cell, skin, eye, nerve, teeth and muscle health, growth, digestibility, absorption, stomach concerns (*e.g.*, "easy on your stomach"), throat, red blood cell, and other miscellaneous health claims. Many of the claims in the nondisease health claims category would be considered "structure-function" claims in FDA terminology.[2]

As shown in Figure 5-1, explicit *disease claims* are not the majority of health claims during this period. Explicit disease claims are made in less than 1 percent of advertisements through 1984 and never in more than 4.6 percent of ads throughout the period.

When affiliated claims are taken together with disease claims, however, the picture is more variable. Disease and affiliated claims do constitute the majority of health claims from 1983 until 1992, and again after 1995. The percentage of advertising that includes these claims is well under 2 percent through 1982. The percentage of ads with these claims rises steadily after 1982, peaking at 8.7 percent of food ads in 1989. The use of these categories of health claims falls precipitously after 1990, to 0.7 percent of ads in 1993, and begins rising again only in 1995. By 1997, 6.3 percent of advertisements include a disease or affiliated claim, approximately 72 percent of the 1989 peak use.

[2] A *structure-function claim* is a term of art in FDA labeling regulation, which is usually described as a claim that explains the role of a nutrient or ingredient in affecting or maintaining the structure or function of some system in humans. See, for instance, "Staking a Claim to Good Health, FDA and Science Stand Behind Health Claims on Foods," Paula Kurtzweil, *FDA Consumer Magazine*, November-December 1998.

Figure 5-1 Percentage of Ads with *Disease and Affiliated Claims* Relative to All *Health Claims*[1]

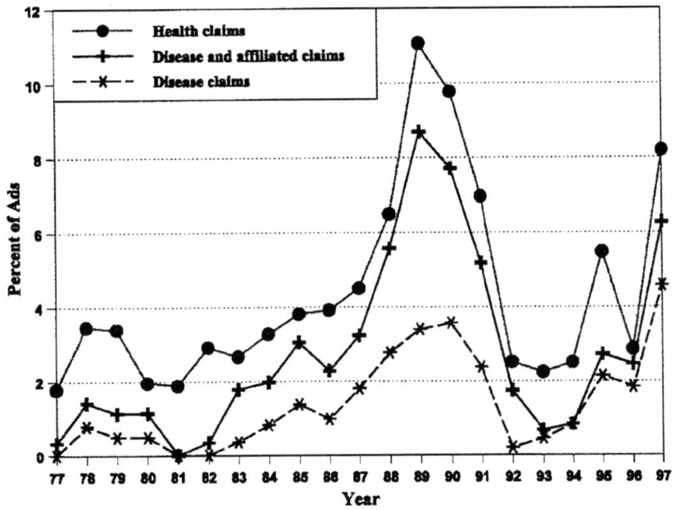

Figure 5-2 Percentage of Ads with *Heart/Serum Cholesterol Claims* Relative to All *Disease and Affiliated Claims*[2]

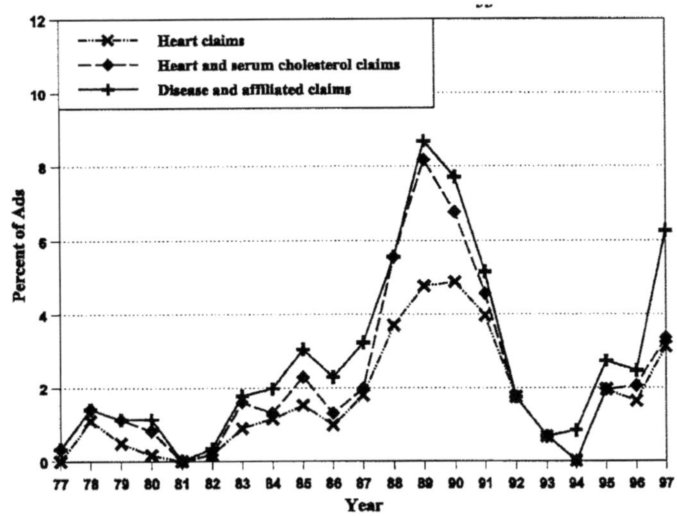

Notes. [1] *Disease claims* specifically mention diseases, while *Affiliated claims* are defined here to include claims closely affiliated with disease claims, such as serum cholesterol claims. *Health claims* are defined to include all health effect claims, including but not limited to *Disease and affiliated claims.*

[2] *Heart claims* include both *heart disease claims* and *other heart claims, e.g., heart smart.*

In the early years of the sample from 1977 to 1983, the majority of health claims are nondisease health claims rather than disease or affiliated claims. In particular, it is worth noting that in the late 1970s and early 1980s, many of these health claims dealt with "digestibility" or "regularity." A substantial portion of this total is due to advertising for Crisco oil and shortening, which uses the tag line "it's digestible" for much of the period. The other nondisease health claims during the period are primarily bone claims for children and teeth claims.

Note also that in the early period and in the post-1990 period, when the use of disease and affiliated claims is very low, the use of nondisease health claims grows relative to disease and affiliated claims (shown by the gap between the top graph and the second graph in Figure 5-1). When explicit disease and affiliated claims are not used, producers appear to shift somewhat to less explicit health claims.

Heart Disease Claims

As shown in Figure 5-2, heart-related claims are the most common type of health claim throughout the years of this study. Claims that specifically mention heart disease or other types of heart claims, *e.g., heart smart* or *for your heart's sake*, rise to 1.1 percent of ads in 19708. After 1983, the use of heart claims rises steadily, increasing its rate of increase after 1987. The use of explicit heart claims peaks at 4.9 percent of all ads in 1990, before falling precipitously in the early 1990s. Heart claims increase again only after 1994, reaching 3.1 percent of ads in 1997.

Serum cholesterol claims have also been a major feature of health claim advertising during the period.[3] The second graph in Figure 5-2 shows the combined use of heart and serum cholesterol claims in food advertising. The pattern is largely the same as for heart claims alone, but the magnitudes are higher. In particular, when compared to the use of *any* disease or affiliated claim (the top line in the figure), heart and serum cholesterol claims together make up the majority of disease or affiliated claims throughout the period.[4]

Together heart and serum cholesterol claims are used a bit in the late 1970s and again beginning in 1983. Use peaks in 1989 at 8.2 percent of ads that year, before falling to 0 percent in 1994, and then rising again to 3.4 percent in 1997. The level and changes in use of heart and serum cholesterol claims are the most dramatic of any category of health claim.

Heart and serum cholesterol claims are made for a wide variety of products, but the frequency of their use varies substantially by product category. Table 5-1 provides data on the use of heart and serum cholesterol claims across the major food categories defined in Table 3-2. Heart and serum cholesterol claims are used most frequently in the Fats & Oils category as competition focuses on the type of fat in the products and why that should be important to consumers.[5] These products are among the first to use heart-related claims as producers attempt to convey the reasons to choose one type of fat over another. At its peak in the late 1980s, nearly 45 percent of all ads in this category make a heart or serum cholesterol claim.

[3] Serum cholesterol claims include claims such as "can help reduce high levels of cholesterol in the blood" or "can help reduce the serum cholesterol that's already there" or "helps lower your cholesterol."

[4] The sample also includes a small number of ads with a picture of a heart but no heart or serum cholesterol claim. Typically these ads include saturated fat or cholesterol content claims. These ads with heart pictures did not occur frequently enough to change the results reported here.

[5] During the period examined here, saturated fats are widely recognized as a significant dietary concern in raising the risk of heart disease (National Research Council, 1989, p. 537). Thus, fats high in polyunsaturated or monounsaturated fats are desirable substitutes for fats high in saturated fats. In the 1990s, scientific evidence also grew to indicate that transfatty acids (part of monounsaturated fats) may also play a role in increasing the risk of heart disease.

Heart and serum cholesterol claims are also relatively common in the Meat/Eggs category. This food category includes egg and meat substitutes, *e.g., Egg Beaters* or vegetable patties, as well as frozen entrees and other prepared foods in which meat is the main ingredient, *e.g., Healthy Choice* meat dinners. Heart and serum cholesterol claims appear primarily in advertising for these named products within the Meat and Eggs category and peak at 22 percent of all advertising in the category in 1990. Heart and serum cholesterol claims appear in 8.6 percent of ads in the category in 1997. To a lesser extent, heart and serum cholesterol claims are also made for frozen entrees in which the main ingredient is poultry, fish or grain during the late 1980s. These do not appear again after 1991.

The Cereals/Breads category is another category in which heart and serum cholesterol claims play a substantial role. These claims arise in the late 1980s and are primarily made for oat bran and other oat cereals. These claims stop in 1990 and resume in 1997.[6] A few other cereals make serum cholesterol claims in October 1995 and 1996 based on their low fat and low cholesterol content.

The table also includes data for the Fruit/Vegetables category, which includes fruit juices. In the late 1980s and early 1990s, heart and serum cholesterol claims are made in a few advertisements for lima beans and pears. Post-1990 heart-related claims are made only in orange and grapefruit juice advertising.

Dairy had a few heart-related claims for low and nonfat milk and yogurt products in the late 1980s and after 1990. Dressings/*etc.* had heart-related claims for low or no fat salad dressings in the early 1990s but are prohibited under current labeling rules. Soft Drinks/*etc.* have no heart or serum cholesterol claims.

Finally, the table includes data for Desserts/Salty Snacks/Sweet Breads, which is a category that includes most dessert, salty snack, donut and related items, as defined in Table 3-2. Heart and serum cholesterol claims are not a significant feature of advertising in this product category at any time during the period.[7]

Cancer Claims

As shown in Figure 5-3, the percentage of advertisements that make a cancer claim is much smaller than that for heart-related claims throughout the years covered here. In 1980, a few soft drink ads contain claims about cancer in laboratory animals consuming artificial sweeteners. With these exceptions, cancer claims begin in the cereal market in 1984 and highlight fiber content. Fruit sellers and juice producers quickly join the cereal producers in making cancer claims, but together these claims never rise above one percent of food ads during the mid and late 1980s before falling to zero percent in 1992 and 1993. After 1993, cancer claims are used again, rising to 2 percent of all food ads by 1997. These post-1990 claims are primarily from juice producers, together with a few cereal company claims.

Table 5-2 provides cancer claim data by product category. Cancer claims are most prominent in the Cereals/Bread category in the mid-1980s but are eliminated in 1991 and reappear only in 1997.[8] The Fruit/Vegetables category is the only other category in which

[6] The FDA authorized health claims relating oat products to coronary heart disease risk on January 23, 1997 (62 *Fed. Reg.* 3584).

[7] The only heart-related ads for this category include two ads for a low fat, low cholesterol muffin mix with the line "as part of a low fat, low cholesterol diet can help reduce cholesterol" and an ad for peanut butter with the line "good nutrition straight from the heart."

[8] These ads may have been spurred by competition with oat cereals making the newly authorized oat-heart disease claims in 1997, as described above.

cancer claims were used with any frequency. These claims were made for a few vegetable and fruit products in the mid-1980s and for juices post-1990.

Figure 5-3 Percentage of Ads with *Cancer Claims*[1]

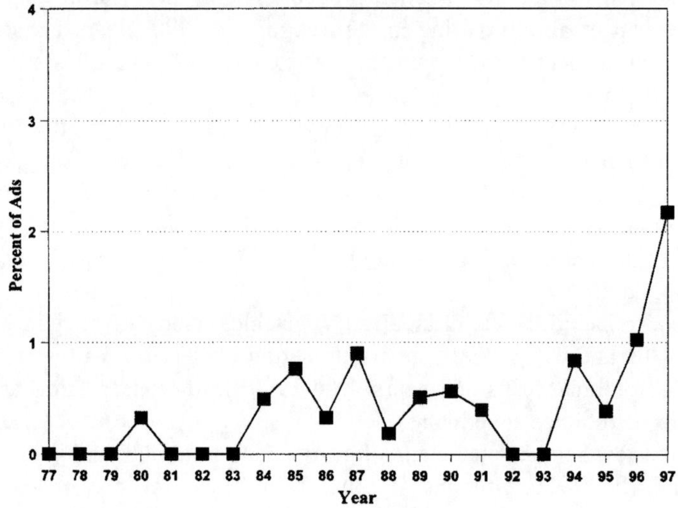

Figure 5-4 Percentage of Ads with *Osteoporosis or Other Bone Claims*[2]

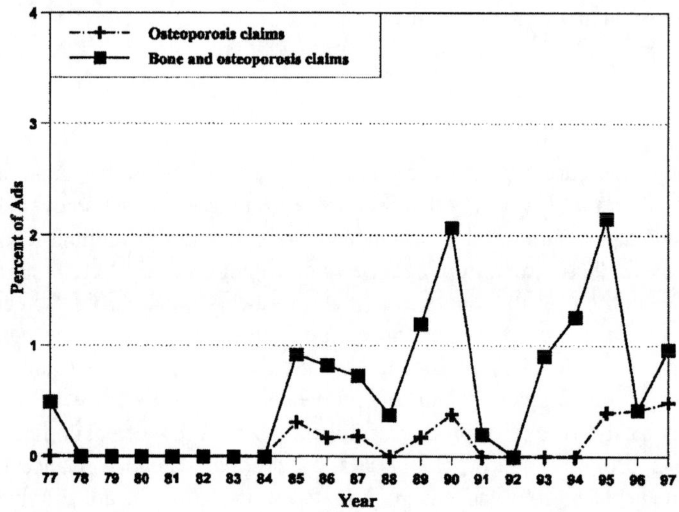

Notes. [1] *Cancer claims* category includes all claims that explicitly mention cancer.

[2] *Osteoporosis claims* includes all claims that explicitly mention osteoporosis. *Bone claims* include any other type of bone claim, *e.g., build strong bones* or *for your bones.*

Table 5-1 Percentage of Ads with *Heart/Serum Cholesterol Claims* for Select Product Categories[1]

Year	All Foods	Meat/Eggs	Poultry/ Fish/Grain	Cereals/ Breads	Dairy	Fats &Oils	Desserts/ Snacks	Fruit/ Vegs.	Drinks	Dressing/ *Etc.*
1977	0.3	0.0	0.0	0.0	0.0	7.1	0.0	0.0	0.0	0.0
1978	1.4	8.3	0.0	0.0	0.0	11.4	0.0	0.0	0.0	0.0
1979	1.1	1.9	1.3	0.0	0.0	10.0	0.0	0.0	0.0	0.0
1980	0.8	2.6	0.0	0.0	0.0	6.5	0.0	0.0	0.0	1.7
1981	0.0	0.0	0.0	0.0	0.0	0.0	0.0	0.0	0.0	0.0
1982	0.2	0.0	1.2	0.0	0.0	0.0	0.0	0.0	0.0	0.0
1983	1.6	0.0	0.0	0.0	0.0	20.5	0.0	0.0	0.0	1.2
1984	1.5	0.0	0.0	0.0	0.0	19.5	0.0	0.0	0.0	0.0
1985	2.3	0.0	0.0	0.0	3.4	21.8	0.5	0.0	0.0	0.0
1986	1.3	3.4	0.0	0.0	1.5	14.3	0.0	0.0	0.0	0.0
1987	2.0	4.3	2.3	1.9	2.4	12.5	0.0	0.0	0.0	1.1
1988	5.6	2.6	1.1	18.6	1.4	45.0	0.0	1.1	0.0	0.0
1989	8.2	11.5	5.3	18.2	1.5	39.0	2.2	0.9	0.0	1.1
1990	6.8	22.2	0.0	26.5	0.0	44.4	0.0	0.0	0.0	1.8
1991	4.6	5.8	2.8	0.0	0.0	36.7	0.0	2.5	0.0	7.8
1992	1.7	2.0	0.0	0.0	0.0	5.4	0.0	0.0	0.0	7.1
1993	0.7	0.0	0.0	0.0	0.0	2.8	0.0	5.1	0.0	1.4
1994	0.0	0.0	0.0	0.0	0.0	0.0	0.0	0.0	0.0	0.0
1995	2.0	0.0	0.0	3.0	8.5	0.0	0.0	0.0	0.0	1.7
1996	2.1	5.9	0.0	4.0	1.9	0.0	0.0	10.0	0.0	0.0
1997	3.4	8.6	0.0	18.2	0.0	0.0	0.0	6.1	0.0	0.0

Notes. [1] Product categories are defined in Table 3-2.

Table 5-2 Percentage of Ads with *Cancer Claims* for Select Product Categories[1]

Year	All Foods	Meat/Eggs	Poultry/ Fish/Grain	Cereals/ Breads	Dairy	Fats &Oils	Desserts/ Snacks	Fruit/ Vegs.	Drinks	Dressing/ *Etc.*
1977	0.0	0.0	0.0	0.0	0.0	0.0	0.0	0.0	0.0	0.0
1978	0.0	0.0	0.0	0.0	0.0	0.0	0.0	0.0	0.0	0.0
1979	0.0	0.0	0.0	0.0	0.0	0.0	0.0	0.0	0.0	0.0
1980	0.3	0.0	0.0	0.0	0.0	0.0	0.0	0.0	3.6	0.0
1981	0.0	0.0	0.0	0.0	0.0	0.0	0.0	0.0	0.0	0.0
1982	0.0	0.0	0.0	0.0	0.0	0.0	0.0	0.0	0.0	0.0
1983	0.0	0.0	0.0	0.0	0.0	0.0	0.0	0.0	0.0	0.0
1984	0.5	0.0	0.0	2.2	0.0	0.0	0.0	2.1	0.0	0.0
1985	0.8	0.0	0.0	1.6	0.0	0.0	0.0	4.0	0.0	0.0
1986	0.3	0.0	0.0	5.3	0.0	0.0	0.0	0.0	0.0	0.0
1987	0.9	0.0	1.1	7.7	0.0	0.0	0.0	0.0	0.0	0.0
1988	0.2	0.0	0.0	2.3	0.0	0.0	0.0	0.0	0.0	0.0
1989	0.5	0.0	0.0	3.6	0.0	0.0	0.0	0.0	0.0	1.1
1990	0.6	0.0	0.0	8.8	0.0	0.0	0.0	0.0	0.0	0.0
1991	0.4	0.0	0.0	9.5	0.0	0.0	0.0	0.0	0.0	0.0
1992	0.0	0.0	0.0	0.0	0.0	0.0	0.0	0.0	0.0	0.0
1993	0.4	0.0	0.0	0.0	0.0	0.0	0.0	0.0	0.0	0.0
1994	0.0	0.0	0.0	0.0	0.0	0.0	0.0	7.7	0.0	0.0
1995	1.4	0.0	0.0	0.0	0.0	0.0	0.0	4.6	0.0	0.0
1996	1.0	0.0	0.0	0.0	0.0	0.0	0.0	10.0	0.0	0.0
1997	2.2	0.0	0.0	9.1	0.0	0.0	1.2	8.2	0.0	0.0

Notes. [1] Product categories are defined in Table 3-2.

Table 5-3 Percentage of Ads with *Osteoporosis/Other Bone Claims* for Select Product Categories [1]

Year	All Foods	Meat/Eggs	Poultry/ Fish/Grain	Cereals/ Breads	Dairy	Fats &Oils	Desserts/ Snacks	Fruit/ Vegs.	Drinks	Dressing/ *Etc.*
1977	0.5	0.0	0.0	4.8	0.0	0.0	0.0	1.0	1.9	0.0
1978	0.0	0.0	0.0	0.0	0.0	0.0	0.0	0.0	0.0	0.0
1979	0.0	0.0	0.0	0.0	0.0	0.0	0.0	0.0	0.0	0.0
1980	0.0	0.0	0.0	0.0	0.0	0.0	0.0	0.0	0.0	0.0
1981	0.0	0.0	0.0	0.0	0.0	0.0	0.0	0.0	0.0	0.0
1982	0.0	0.0	0.0	0.0	0.0	0.0	0.0	0.0	0.0	0.0
1983	0.0	0.0	0.0	0.0	0.0	0.0	0.0	0.0	0.0	0.0
1984	0.0	0.0	0.0	0.0	0.0	0.0	0.0	0.0	0.0	0.0
1985	0.9	0.0	0.0	0.0	6.9	0.0	0.0	0.0	6.9	0.0
1986	0.8	0.0	0.0	0.0	5.9	0.0	0.0	0.0	2.9	0.0
1987	0.7	0.0	0.0	1.9	9.5	0.0	0.9	0.0	0.0	0.0
1988	0.4	0.0	0.0	2.3	0.0	0.0	0.0	1.1	0.0	0.0
1989	1.2	0.0	0.0	0.0	0.0	0.0	0.0	6.4	0.0	0.0
1990	2.1	0.0	0.0	0.0	0.0	0.0	0.0	12.6	0.0	0.0
1991	0.2	0.0	0.0	0.0	0.0	0.0	0.0	1.2	0.0	0.0
1992	0.0	0.0	0.0	0.0	0.0	0.0	0.0	0.0	0.0	0.0
1993	0.9	0.0	0.0	0.0	2.9	0.0	0.0	0.0	10.3	0.0
1994	1.3	0.0	0.0	0.0	0.0	0.0	0.0	0.0	23.1	0.0
1995	2.1	0.0	0.0	0.0	8.5	0.0	0.0	9.3	0.0	0.0
1996	0.4	0.0	0.0	0.0	3.8	0.0	0.0	0.0	0.0	0.0
1997	1.0	0.0	0.0	2.3	5.2	0.0	0.0	0.0	0.0	0.0

Notes. [1] Product categories are defined in Table 3-2.

Osteoporosis or Other Bone Claims

Claims about the role of calcium in preventing osteoporosis have also been used in advertising.

As shown in Figure 5-4, claims that mention osteoporosis explicitly are quite limited throughout the period examined here, never rising above 0.5 percent of advertising. These claims are primarily for dairy products, cereals served with milk, and fruit juices fortified with calcium. As with the health claims described above, these claims begin in the mid-1980s, disappear by 1991, and resume in 1995.

Figure 5-4 also shows the use of other types of bone claims, together with explicit osteoporosis claims. These bone claims follow a similar pattern. With the exception of a few cereal claims in 1977, bone claims do not begin until after 1984. Their use falls dramatically in 1991, before rising again in 1993. By 1997, bone and osteoporosis claims are found in just under one percent of all ads in the sample. The post-1990 osteoporosis claims are restricted to milk products.

As shown in Table 5-3, bone and osteoporosis claims are largely limited to dairy products, calcium-fortified fruit juices and fruit-flavored drinks, and to cereals fortified with calcium or advertised for use with milk.

Hypertension Claims

Claims about hypertension, high blood pressure, or stroke are coded together as a class. As shown in Figure 5-5, prior to 1990, advertising that mentions hypertension-related issues peaks in 1984 at 1.0 percent of all food advertising. The claims disappear in 1990 and do not return until 1995, when they occur in 1.2 percent of all ads. Prior to 1990, these claims are made in ads for grapefruit products, corn oil, milk, fish, and meat substitute products. After 1990, these claims appear only in ads for milk, orange juice, and cereal products.

Birth Defect Claims

Claims discussing the role of folic acid in preventing birth defects are a post-1990 phenomenon. Approximately 0.8 percent of ads have such a claim in 1994 and 1996. These claims are exclusively in orange juice advertising.

Figure 5-5 Percentage of Ads with *Hypertension Claims*[1]

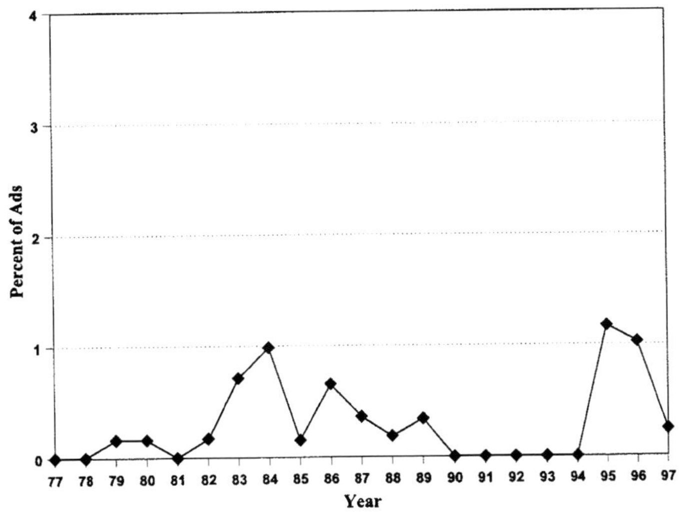

Figure 5-6 Percentage of Ads with *Birth Defect* or *Diabetes Claims*[2]

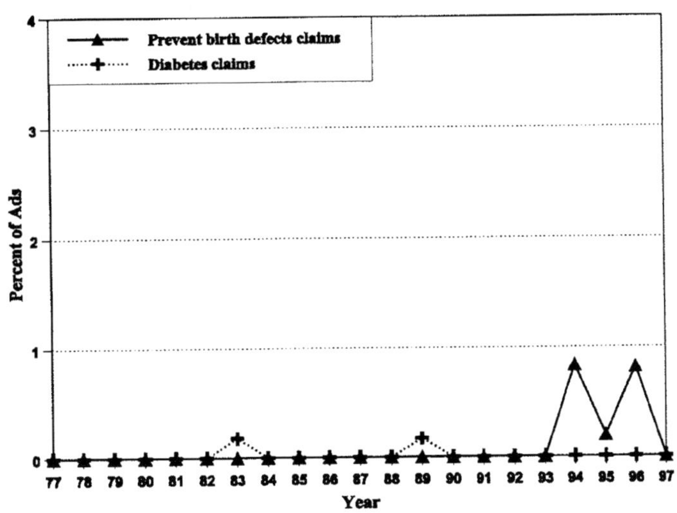

Notes. [1] *Hypertension claims* category includes all claims about hypertension, blood pressure or stroke.

[2] *Birth defects claims* category includes all claims that mention prevention of birth defects. *Diabetes claims* category include all claims that mention diabetes, good for diabetics, *etc*.

Diabetes Claims and Cell Damage/Oxidization/Free Radical Claims

Coders were instructed to categorize any claims in these two categories. No advertisements in our sample have cell damage, oxidization, or free radical claims. Only two advertisements have diabetes claims.

Tooth Decay and Other Tooth Health Claims

As shown in Figure 5-7, claims about not promoting tooth decay or about building strong teeth have been a part of food advertising during most years of our study. These claims are typically used in advertising for sugar-free products to point out their tooth decay advantage relative to sugared products, or they are used for calcium-containing products to highlight the role of calcium in building strong teeth.

Some tooth-decay claims are made in the late 1970s and early 1980s, but these claims increase in 1984 and in the later 1980s to nearly 1 percent of all ads, before disappearing in 1991. The only post-1990 claims related to teeth are a few claims about calcium's role in producing stronger teeth for calcium-fortified juice and milk products in 1995.

Regularity Claims

Coders were also instructed to code claims about regularity, keeping the digestive system functioning, or related claims. The use of these claims is also shown in Figure 5-7. Though not a large feature of advertising, these claims are used periodically to highlight one of the benefits of fiber consumption. In particular, in a 1982 precursor to their 1984 cancer campaign, the Kellogg company began an advertising campaign on the benefits to consumers of bran cereals to "keep their digestive system functioning smoothly." These claims were also used by a few other producers during the late 1980s fiber-cancer period, and then again by Kellogg in 1992 and 1993 when they did not use cancer claims.

All Other Health Claims

Finally, Figure 5-8 illustrates the use of all other health claims during this period. These claims include a wide variety of other health effect claims (but no disease claims). The data shows no systematic trend over the period and never rises above 2 percent of all ads. Many of the advertisements that have a claim in this category also have another health claim coded in one of the above categories. Thus, these data indicate that our coding scheme has captured most health claims in one of our identified subcategories of health claims.

Figure 5-7 Percentage of Ads with *Tooth Decay/Tooth Health* or *Regularity Claims*[1]

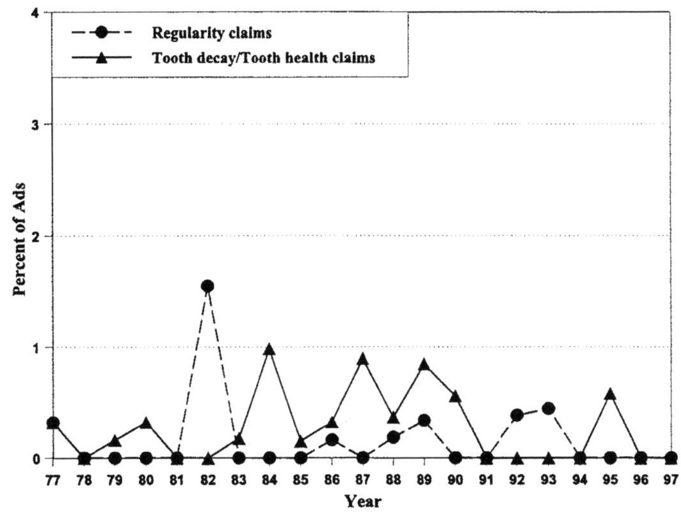

Figure 5-8 Percentage of Ads with *Other Health Claims*[2]

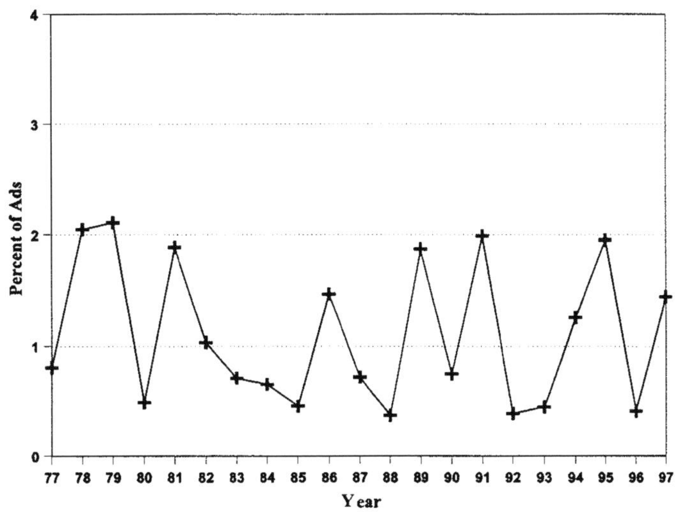

Notes. [1] *Tooth decay/Tooth health claims* category includes all claims about preventing tooth decay, maintaining strong teeth, or other aspects of tooth health. *Regularity claims* category includes all regularity, digestive system or related claims.

[2] *Other health claims* includes all health effects claims not covered elsewhere. See text.

One substantial subcategory within this *Other Health Claims* group consists of claims about "digestibility" or other "easy on the stomach" claims. Approximately 35 percent of all claims in this miscellaneous category are digestibility claims. These claims are made for fat or oil products, lactose-reduced dairy products, acid-reduced juice products, soups, and yogurt. Other claims deal with the role of iron and folic acid in red blood cell development, protein for muscles, vitamins or minerals that help absorption of other nutrients, and other miscellaneous claims. Many of the claims in this category would be considered "structure-function" claims in FDA terminology.

Conclusion

The evidence on the use of health claims indicates that advertisers increased their use of health claims in the mid-1980s. The evidence also indicates that the use of health claims was sharply diminished in the early 1990s but that by 1997 the use of health claims had recovered to 74 percent of the peak 1989 level of use. In 1997 disease and affiliated claims are at 72 percent of their 1989 level, and disease claims are at 128 percent of their 1989 peak.

Heart and serum cholesterol claims are the most frequently used health claims both before and after 1990, but the use of these heart-related claims also diminishes most in the post-1990 years. In 1997 heart claims are at 64 percent of their 1989 peak, and heart and serum cholesterol claims together are at 41 percent of their 1989 peak. In large part these reductions appear to be due to the elimination of heart-related claims for fat and oil products, and to a lesser extent, to the elimination of these claims for low and nonfat salad dressings.

Cancer claims are the health claims that have increased most post-1990, primarily due to the use of these claims by juice producers. Folic acid-birth defect claims are also a post-1990 phenomenon, but they remain at a low level of use.

Chapter 6

REGULATION AND ADVERTISING CLAIMS

INTRODUCTION

Nutrition-related claims in marketing have been the subject of considerable regulatory scrutiny during the years 1977 to 1997. Claims on food labels are primarily regulated by the Food and Drug Administration (FDA),[1] and claims in food advertising are under the primary jurisdiction of the Federal Trade Commission (FTC). Both agencies initiated major rulemakings to change the rules governing nutrition-related claims on labels and in advertising during the years of our sample. These regulatory changes have been described in detail in a number of previous publications,[2] but they represent an important backdrop against which producers make decisions about which claims to make in their advertising.

In this chapter, we briefly review some of the key regulatory events that could affect food producers' incentives to make nutrition-related claims in their advertising. We also present analyses of the timing of these regulatory changes and the use of different types of nutrition-related claims. Finally, we focus in more depth on some of the substantial changes that have occurred since the implementation of the NLEA rules in the mid-1990s.

KEY REGULATORY EVENTS

Table 6-1 summarizes the key regulatory events that could affect the use of nutrition-related claims in advertising. Note that on December 31, 1974 nutrition labeling is required for foods with added nutrients, and for foods whose labels or advertising include nutrition-related claims.[3] Producers are also free to voluntarily label their other products with the standard nutrition label. In May 1994, the regulations issued under the *Nutrition, Labeling and Education Act of 1990* (NLEA) make the nutrition label mandatory for virtually all packaged food products. So throughout the years 1977 to 1997, foods making nutrition-related claims of any type must have nutrition labels on their packaging; after May 1994, virtually all foods have such labels. These labeling requirements establish standard methods for measuring major nutrients.

---

[1] Labels for meat and poultry products are regulated by the U. S. Department of Agriculture (USDA).

[2] See, for instance, Beales and Muris (1993), Ippolito and Mathios (1996), and Pappalardo and Ringold (2000).

[3] Legislative and regulatory citations are listed in Table 6-1.

Advertising rules never formally prohibit or explicitly regulate nutrition-related claims, including general nutrition claims, nutrient content claims, or health claims, as those terms are used here. All advertising claims are subject to general advertising enforcement under the FTC's authority to pursue deceptive business practices.[4] An assessment of what the agency considers to be deceptive, and thus the enforcement risk attached to different types of claims, must be inferred from cases or other agency pronouncements during the period of interest.

A variety of evidence suggests that health claims and some nutrient content claims raise substantial legal risk at the FTC during the period from the mid-1970s to at least early 1983. The same is true for general nutrition claims and other nutrient content claims from the mid-1970s until late 1980. These judgments are based in part on key events in the FTC's Food Rulemaking listed in Table 6-1.

The FTC Food Rulemaking was a broad effort to regulate food claims in advertising through explicit industry-wide rules that began with the publication of the staff's initial regulatory proposal on November 11, 1974, well before the start of the period examined here. The initial proposal has several major components, including a ban on all diet-disease claims as inherently deceptive, and plans to regulate general nutrition claims, "emphatic" nutrient content claims, such as "loaded with" or "high in," and several specific nutrient content claims. On March 17, 1978, the Presiding Officer's report in the rulemaking specifically proposes to define rules to allow fatty acid/heart disease claims, thus recommending against a ban on diet-disease claims, but not clarifying the conditions for such health claims.

After extensive hearings and considerable controversy, on April 8, 1980, the FTC votes to end Part II of the Food Rule, which would have regulated general nutrition terms, such as "nutritious" or "health food," and "emphatic" nutrient claims, preferring to deal with these issues on a case-by-case basis under its general authority. The FTC, however, does not end the rulemaking regarding fatty acid/heart disease claims, some other nutrient content claims, and other general nutrition claims, such as "natural." After further consideration, the FTC votes to close the entire Food Rulemaking on December 17, 1982 in favor of case-by-case enforcement. This ends proposals to explicitly regulate fat and cholesterol claims, calorie and weight claims, "natural" claims, and fatty acid/heart disease claims in advertising. This decision is finalized after a period of public comment in Spring 1983.

Thus, by Spring 1983 at the latest, the FTC makes it clear that nondeceptive health claims, general nutrition claims, and nutrient content claims in advertising face considerably less enforcement risk at the FTC.

---

[4] A listing of FTC food advertising cases during these years is provided in Appendix C.

Table 6-1 Key Regulatory Events Regarding Nutrition and Health Claims

Date	Event	Predicted Effects of Key Events on Claim Use
November 11, 1974	FTC staff proposes Food Rule regulating general health and other nutrient claims in ads; would prohibit health claims.[1]	
December 31, 1974	FDA nutrition labeling rules in effect requiring nutrition label for many foods.[2]	
March 17, 1978	Presiding Officer's report in FTC rulemaking proposes fat/cholesterol heart disease claims.[3]	
* *April 8, 1980*	FTC votes to end Part II of Food Rulemaking; would have regulated general nutrition claims and emphatic nutrient claims.[4]	+ for general claims - for health claims ? for nutrient claims
* *December 17, 1982*	FTC votes to end entire Food Rulemaking, including energy, weight control, fatty acid, heart disease, and natural claims. Policy reverts to deception/substantiation standards for claims.[5]	+ for health claims + for nutrient claims + for general claims
October 7, 1984	Kellogg fiber/cancer advertising campaign begins. Not challenged by FDA or FTC.[6]	
* *August 4, 1987*	FDA publishes proposed label rule for health claims based on an *ex post* deception standard.[7]	+ for health claims + for nutrient claims
* *February 13, 1990*	FDA withdraws 1987 health claim proposal; announces plan for more restrictive regulations.[8]	- for health claims - for nutrient claims
July 19, 1990	FDA proposes extensive labeling rules; mandatory nutrition label, standardized claims, prohibition of unapproved claims.[9]	
November 8, 1990	President signs NLEA authorizing mandatory nutrition label and nutrition claim label rules; sets up process for regulating health claims on labels.[10]	
November 27, 1991	FDA proposes new nutrient content and general health claim rules under NLEA; also proposes preapproval system for label health claims under "significant scientific agreement" standard.[11]	

(Continued next page.)

Table 6-1 (Continued)

Date	Event	Predicted Effects of Key Events on Claim Use
May 8, 1993	FDA final NLEA label rules effective for health claims. Model claims for a limited number of preapproved claims and foods.[12]	
* *May 8, 1994*	FDA final NLEA label rules effective for nutrition claims.[13]	? for nutrient claims - for comparative claims + for health claims
* *May 13, 1994*	*FTC Enforcement Policy Statement on Food Advertising* coordinates advertising policy with labeling policy.[13]	
December 21, 1995	FDA publishes proposal to simplify health claim rules to allow shorter claims for more foods.[14]	

Notes. * indicates key regulatory events used in regression estimates. FDA is Food and Drug Administration; FTC is Federal Trade Commission; NLEA is *Nutrition, Labeling and Education Act of 1990.*

[1] ***Federal Register***, 39, November 11, 1974, 39812.

[2] ***Federal Register***, 38, March 14, 1973, 6951. Nutrition labeling is required on food containing added nutrients or whose label or advertising includes nutrition-related claims.

[3] Dixon (1978) or see Gordon, Richard L., "FTC on food; 'Health' is out; calories count," *Advertising Age*, March 20, 1978, 1.

[4] ***Federal Register***, 45, April 8, 1980, 23705. Part II of the rule would have regulated "emphatic" nutrition claims, such as "lots of," "high in," "packed with," "excellent source of," *etc.*, general health claims about the nutritional value of a food, such as "nutritious," "wholesome," *etc.*, and nutrient content claims in general.

[5] ***Federal Register***, 48, May 24, 1983, 23270. This ended the remaining portions of the Food Rulemaking, which would have regulated energy and calorie claims, fat, fatty acid, and cholesterol claims, natural and organic claims, and health and related claims. Claims continued to be subject to substantiation and deception standards of general advertising policy.

[6] Colford, Steven W., "Kellogg eyes long run for All-Bran ads," *Advertising Age*, January 7, 1985, 64.

[7] ***Federal Register***, 52, August 4, 1987, 28843; this proposal would formally end the prohibition of all specific health claims on labels.

[8] ***Federal Register***, 55, February 13, 1990, 5176.

[9] ***Federal Register***, 55, July 19, 1990, 29487.

[10] ***Public Law*** No. 101-535, 104 Stat. 2353 (codified in part at 21 U.S.C. 343(i), (q) and (r)). Because of the proximity of the 1990 events, we will effectively combine them in the regression tests.

[11] ***Federal Register***, 56, November 27, 1991, 60365, 60537.

[12] ***Federal Register***, 58, January 6, 1993, 2478.

[13] ***Federal Register***, 59, June 1, 1994, 28388.

[14] ***Federal Register***, 60, December 21, 1995, 66206.

During the early years of the period examined here, health claims in advertising also raise the risk of legal challenge by the FDA.[5] During these years, a health claim in advertising allows the FDA to declare the product a "drug," and thus, subject to drug law regulations.[6] Food products do not have the type of efficacy testing required for drug products, thus under this interpretation, the use of health claims in advertising raises substantial legal risk at the FDA.

Two events are listed in Table 6-1 as key points when producers are likely to perceive the risk of FDA prosecution of health claims to be reduced. On October 7, 1984, the Kellogg company begins a highly publicized TV advertising campaign explicitly using the National Cancer Institute's statements on the potential relationship between fiber and cancer to promote its high fiber cereals. Despite much public discussion and considerable controversy, neither the FDA nor the FTC files charges against Kellogg. The second event occurs on August 4, 1987, when the FDA publishes its long awaited proposed rule to govern health claims, which significantly relaxes its prior position on such claims.[7]

Thus, by August 1987, and possibly earlier, the regulatory risk from making nondeceptive diet-disease claims in advertising is substantially less than in the early years of our sample. This decision is broadly perceived as opening up the opportunities for food producers to make health claims. The environment again changes in early 1990, when the FDA withdraws its 1987 proposal regarding health claims, and in July 1990, proposes an extensive revision of its labeling rules to include mandatory nutrition labeling, standardized nutrient content and general nutrition claims, a more stringent preapproval system for health claims, and a prohibition of all unauthorized nutrition or health claims.

This is followed in November 1990 by the passage of the NLEA, which clarifies FDA's authority to set up rules that govern health claims, nutrition labeling, and nutrition claims of all types on labels. Key events in the NLEA implementation process are listed in Table 6-1 and include November 27, 1991, when the FDA reproposes its 1990 labeling rules, with some adjustments required by the NLEA. Thus, through a series of events in 1990 and 1991, the rules governing all types of nutritionrelated claims are revisited, and the process of completing final rules under the NLEA is initiated.

The final NLEA label rules for health claims are effective in May 1993 and for nutrient content claims and general nutrition claims in May 1994. Also in May 1994, the FTC issues an enforcement policy statement harmonizing advertising policy with the new label requirements for all types of nutrition-related claims. Key features of the NLEA-based label rules include a listing of approved terms, explicit requirements for nutrient content claims, triggered disclosures in some cases, *e.g.,* for comparative claims, and explicit requirements for some general nutrition claims, such as the term "healthy." The NLEA-based rules also provide for only a limited number of health claims, and put specific restrictions on which foods can make such claims.[8]

---

[5] For a more detailed discussion of this regulatory history, see Ippolito and Mathios (1996) 19-26.

[6] See Hutt (1986) 25 or "FDA May Strengthen Ban on Cholesterol-Reduction Claims," *Food Chemical News*, January 25, 1971, 21-22.

[7] The proposed regulations are based on a reasonable basis, deception-type standard. This approach would allow health claims without preclearance from the government and would hold them to a test that they do not mislead consumers and that they have a "reasonable basis" of scientific support. A "reasonable basis" standard for scientific support is generally considered to be a high standard of scientific support, but not necessarily one in which consensus has been reached.

[8] See Ippolito and Mathios (1993) for a more detailed discussion of these rules.

This brief summary of regulatory and legislative events illustrates the substantial changes that occur during the period of our sample. Rules have been relaxed in some cases and restricted in other cases, making it difficult to characterize the changes overall. To get a better assessment of whether changes in the rules appear to be related to changes in the use of particular types of nutrition-related claims, we turn to regression estimates first for health claims, and then for the various classes of nutrition claims.

HEALTH CLAIMS AND REGULATION

Background

The regulatory changes described above are most stark for health claims. The events in the FTC's rulemaking suggest that the 1980 decision not to close the health claim portion of the rulemaking could have been perceived as a negative event for advertisers interested in using health claims, since it follows the 1978 Presiding Officer's report suggesting that health claims for heart disease would be allowed in advertising. In contrast, the December 1982 decision makes it clear that the FTC is willing to allow nondeceptive health claims, and thus, increases the likelihood that producers would make health claims in advertising.

In the early 1980s the implication of the FDA's rules means that advertisers using health claims, especially disease and affiliated claims, face considerable risk from the FDA, so the FTC's rulings may not make much difference in advertisers' assessment of the risk inherent in using health claims. If so, the FDA's 1987 proposal to allow health claims under a standard similar to the FTC's reasonable basis/deception standard should have been a positive event for advertisers, removing the remaining regulatory concern in using nondeceptive health claims.

These major regulatory events for health claims are indicated by asterisks in Table 6-1. The table also includes a prediction for whether the regulatory change should increase or decrease the use of health claims. Note, in particular, that we focus on the FTC's enforcement statement as the key post-NLEA event, but the results do not change significantly if we use the May 1993 effective date for the FDA health claims rules instead.

Methodology

To assess the relationship between the regulatory events and the use of health claims, we estimate various simple regressions that relate the likelihood that an advertiser uses health claims against time and the key regulatory events. For instance, consider the simple linear regression:

$$\begin{aligned} Y_{it} = {} & C + b_1 D_{5/1980} + b_2 D_{1/1983} + b_3 D_{8/1987} + b_4 D_{2/1990} + b_5 D_{5/1994} \\ & + \text{Time} + c_1 \text{Time}\, D_{5/1980} + c_2 \text{Time}\, D_{1/1983} \\ & + c_3 \text{Time}\, D_{8/1987} + c_4 \text{Time}\, D_{2/1990} + c_5 \text{Time}\, D_{5/1994}, \end{aligned} \tag{6-1}$$

where Y_{it} = 1 if advertisement i at time t has the claim under study, 0 otherwise,
Time = Date (in years) - 1900 (so February 1977 is 77.2, *etc.*),

$D_{m/y}$ = if the date of the ad is after month m in year y, (so $D_{5/1980}$ = 1 after May 1980), 0 otherwise.

In this specification, the dummy variables represent the major regulatory events indicated in Table 6-1.

Table 6-2 Regression Results for *Disease and Affiliated Claims* Across Regulatory Periods

Variable	Linear-1	Linear-2	Probit-1	Probit-2
Constant	0.010**	-0.224	-2.323**	-11.536
	(2.85)	(-0.76)	(-28.47)	(-1.62)
$D_{5/1980}$	-0.006	-0.007	-0.343**	-0.241
(FTC ends Part II of Food Rule)	(-1.16)	(-0.68)	(-2.15)	(-0.85)
$D_{1/1983}$	0.020**	0.017*	0.682**	0.907**
(FTC ends entire Food Rule)	(3.88)	(1.66)	(4.66)	(2.58)
$D_{8/1987}$	0.050**	0.007	0.532**	0.140
(FDA health claim proposal)				
$D_{2/1990}$	-0.044**	-0.034**	-0.443**	-0.041
(FDA withdraws 1987 proposal/NLEA)	(-8.01)	(-3.22)	(-5.93)	(-0.31)
$D_{5/1994}$	-0.003	0.021**	0.048	0.627**
(FTC policy statement; FDA/NLEA rules effective)	(0.63)	(2.01)	(0.59)	(2.67)
Time	—	0.003	—	0.117
		(0.80)		(1.30)
Time*$D_{5/1980}$	—	-0.006	—	-0.346*
		(-0.86)		(-1.67)
Time*$D_{1/1983}$	—	0.005	—	0.277
		(0.91)		(1.46)
Time*$D_{8/1987}$	—	0.025**	—	0.152**
		(4.28)		(2.04)
Time*$D_{2/1990}$	—	-0.048**	—	-0.578**
		(-7.77)		(-6.42)
Time*$D_{5/1994}$	—	0.035**	—	0.582**
		(7.31)		(6.66)
Adj. R-squared	0.015	0.022		
Log-Likelihood			-1371.26	-1335.86
n	11,647			
Mean Dependent Value	0.027			

Notes. t-statistics in parentheses. Dependent variable is equal to 1 if an ad has a disease or affiliated claim, 0 otherwise. ** indicates significance at the 5 percent level, * at the 10 percent level in a two-tailed test.

Results

Table 6-2 presents ordinary least squares regression results for the use of *Disease and Affiliated Claims* for this equation (Linear-2), as well as for a simpler version in which the trend and interaction terms are not included (Linear-1).[9] The table also includes probit versions of these regressions (Probit-1 and Probit-2).[10] These models are estimated from the individual advertising data, that is, using the 11,647 advertisements in the sample.

Also as expected, coefficients on the second FTC event in early 1983, when the FTC ends the entire rulemaking, are positive and statistically significant. Once it is clear that health claims will be judged according to standard deception criteria at the agency, advertisers increase their use of health claims. This result is statistically significant in all formulations. Use of health claims remains low (in the Linear-1 formulation, approximately 3 percent), but is more than double earlier levels.

The use of health claims also increases following August 1987, when the FDA publishes its proposal explicitly liberalizing its policy on health claims. The coefficient on $D_{8/87}$ is positive and significant in both the Linear-1 and Probit-1 specifications, as are the coefficients on the interaction terms in the expanded models. Note also that the size of the coefficient in the Linear-1 model is considerably larger that the earlier FTC event (.05 versus .02), evidence consistent with the view that the FDA rules for labels have an effect on the use of health claims in advertising.

Also as predicted, the use of claims falls significantly following the FDA decision in February 1990 to pull back its 1987 health claims proposal in favor of more restrictive rules.[11] The coefficients are large and highly significant in all specifications either for the direct effect or for the interaction term at that date. Despite the fact that the FTC has not changed its advertising policy at this point, most of the increase in the use of health claims following the FDA's 1987 decision is reversed.

After the FTC harmonization statement in May 1994 and the FDA's implementation of final rules under the NLEA, the use of health claims shows a significant increase in the models that allow both a discrete effect and an interaction effect (Linear-2 and Probit-2) but not in the simpler models. In part, this reflects the fact that most of the increase in the use of health claims in the post-1994 period comes in the last two years of the period. This pattern suggests that the FDA's December 1995 proposal to simplify the rules for making health claims may have been important to advertisers. The FDA proposal, which has not been finalized (as of April 2002), makes it clear that the long and rather complicated model statements in the original health claim regulations are not required and makes additional simplifications in the rules.

Finally, note that controlling for the regulatory events, the time trend is not significant in the estimates.[12]

Taken together, these results are consistent with the view that advertisers respond to the regulatory rules they face in making health claims, and that changes in the regulatory rules

---

[9] We focus on *Disease and Affiliated Claims* because the regulatory shifts are most directly applicable to his subcategory of health claims and most health claims are of this type. Regression results for all health claims are comparable.

[10] Probit regressions are specifically designed to deal with discrete outcome data; in our case, either an advertisement has a particular type of claim ($Y^{it} = 1$) or it does not ($Y^{it} = 0$). Linear ordinary least squares regression models are not constrained to lie between 0 and 1, and thus, are not ideal for use with discrete data. In many cases, however, linear methods give comparable results to discrete models like the probit model. Linear models have the advantage that their coefficients are easily interpreted. When the results are comparable under the two techniques, we will focus on the linear estimates for this reason. See Greene (1977) 874 for a discussion.

[11] Estimates based on the different regulatory dates in 1990 produce comparable results.

[12] The same result holds if Time is added to the simple models (Linear-1 and Probit-1).

can lead to significant changes in the types of claims producers use in their advertising. The easing of the health claim rules in the mid-1980s, in particular, has a substantial and significant effect on whether producers focus on diet-disease issues in their advertising.

NUTRIENT CONTENT CLAIMS AND REGULATION

Background and Method

The regulatory events identified in Table 6-1 could also affect producers' incentives to use nutrient content claims, both directly because the rules govern nutrient claims, and indirectly because the rules affect health claims, and nutrient content claims often accompany health claims.[13] In this section we examine whether there are systematic movements in the use of nutrient content claims for the major nutrients across the different regulatory periods.

In addition to the effects related to health claims described in the previous section, several of the regulatory events have direct implications for the use of nutrient content claims. In May 1980 the FTC decision to end Part II of the Food Rule terminates efforts to regulate "emphatic" nutrient claims, such as "high in fiber" or "lots of calcium," and thus, should increase these types of nutrient claims. On the other hand, the decision continues efforts to regulate energy, fat, and other nutrient claims, and thus could reduce the use of these nutrient claims. Together these effects provide no clear prediction for a change in the use of nutrient content claims at this point.

In contrast, the 1983 formal end to the FTC Food Rulemaking would be expected to increase the use of nutrient content claims, because the direct regulatory concern has been lifted, and because of the expected increase in health claims. The FDA's August 1987 proposal on health claims should also increase the use of nutrient content claims that accompany health claims. The opposite should be true for the 1990 events, which reverse the 1987 proposal and initiate rulemakings to regulate nutrient content claims directly.

Finally, the 1994 NLEA-based rules could increase nutrient content claims compared to their level at that point, because the rules make clear what is allowed for nutrient claims after a period of considerable controversy, and because they explicitly authorize some health claims. On the other hand, the rules impose strict limits on some nutrient claims, and in the case of comparative claims, require more extensive disclosures. Thus, the prediction for the change in the use of nutrient content claims in 1994 is not clear, and in particular, there are substantial reasons to predict that the use of comparative claims may fall. These predictions are noted in Table 6-1.

[13] The evidence indicates that disease claims are usually accompanied by related nutrient claims. For instance, a margarine ad from 1988 is typical: "Zero cholesterol, low saturated fats, and a downright terrific taste. Medical studies show that a diet low in saturated fats and cholesterol can reduce the level of cholesterol already in your body. And that can help reduce your risk of heart disease. ..."

Overall Results for Nutrient Content Claims: Focus Shifts to Total Fat Under NLEA Rules

Table 6-3 presents simple linear regression results (comparable to those for Linear-1 model for health claims in Table 6-2) relating the use of nutrient content claims for each of the major nutrients to the major regulatory changes during these years.[14]

After the first FTC decision in May 1980, 3 of the 8 coefficients are significant and 2 of the 3 are positive, indicating only limited change in the use of nutrient claims across the 8 listed nutrients. In contrast, after the end of the Food Rule in early 1983 and after the FDA's 1987 health claim proposal, the use of nutrient content claims rises considerably, and these increases are widespread across nutrients; 5 of the 8 coefficients in the post-1983 period and 6 of the 8 in the post-1987 period are significant, and all of these significant coefficients are positive.

In the post-NLEA periods, the growth in the use of content claims slows and then reverses. After 1990, when the FDA withdraws its 1987 proposal and initiates rulemakings on nutrient claims, 5 of the 8 coefficients are significant, but only 3 of these are positive. After 1994 when the NLEA-based rules are final, 6 of the 8 coefficients are significant but only 2 of the 6 are positive. Fat, and to a lesser extent calcium, are the nutrients where content claims continue to grow in the post-NLEA period. In contrast, producers reduce their focus on saturated fat, cholesterol, sodium, and calories under the NLEA rules. These changes are all statistically significant.

Comparative Claims Fall under NLEA Rules

Comparative claims are a subset of nutrient content claims. Under the NLEA rules, comparative claims are required to meet a number of specific restrictions and to disclose more information as part of the claim. In particular, producers are required to disclose the comparison product, the percentage (or fraction) that the nutrient is reduced, and the actual amount of the nutrient for both the product and the comparison food.[15] These added disclosures in the NLEA rules increase the cost of making comparative claims. The NLEA rules also place additional constraints on allowed comparisons. Products must have at least 25% less (or more) of the nutrient in question. Products can only be compared to allowed reference foods, as defined by regulation,[16] and the reference food cannot already have a low (or high) level of the nutrient. While advertisers are not directly bound by the FDA rules, FTC policy guidance states that claims not in compliance with the FDA rules would receive careful scrutiny from the FTC.[17]

[14] Probit regressions give comparable results and are not reported.

[15] In the labeling rules, the first two pieces of information must be immediately adjacent to the claim, but the actual amounts of the nutrient may be adjacent to the most prominent claim or on the same panel as the nutrition label. (21 CFR 101.13(j)(2)(ii and iv)). Thus, under the NLEA regulations, a claim of "less fat" would become "25% less fat than our regular product, 8 grams of fat per ounce versus 11 grams per ounce."

[16] For instance, for "less" and "more" claims the regulations allow comparisons only to foods in the same product category (21 CFR 101.13 (j) (1) (i) (A)).

[17] The FTC's *Enforcement Policy Statement on Food Advertising* issued in May 1994 summarizes the agency position on comparative claims as follows:

To assess the use of comparative claims more systematically, we estimate both linear ordinary least squares and probit regressions relating the use of comparative claims to the major regulatory events. The linear results are presented in Table 6-4.[18] The use of comparative claims rises significantly in the pre-NLEA period. Six of 8 coefficients are significant in both the post-1983 period and the post-1987 period, and 11 of the 12 significant coefficients are positive. These results suggest increases in direct competition on the nutritional features of foods.

Following the NLEA, the trends change markedly. In the post-1990 period, 5 of 8 coefficients are significant and in the post-1994 period 6 of the 8 are significant, but only 3 of these 11 significant coefficients are positive. Most notably, in the post-1994 period when the NLEA rules are final, 7 of the 8 coefficients are negative (6 are significant); and the only exception is for fat, which exhibits a small, insignificant rise.

Thus, one of the most consistent changes in food advertising observed in the post-NLEA period is the systematic reduction in the use of comparative nutrient content claims. With the exception of fat, the use of comparative claims is lower for all of the major nutrients in the post-NLEA period relative to the years preceding its passage. In fact, as shown in Figures 4-2, 4-3, and 4-6 through 4-9, comparative claims are virtually eliminated by 1997 for all nutrients except fat.

The data here do not allow us to determine why comparative claims fall so consistently under the NLEA rules. Possibly claims from the earlier period do not meet the minimum requirements for the use of comparative claims under the NLEA (*e.g.*, the required minimum 25 percent reduction), and advertisers believe that the risk of continuing to make these claims in advertising is too great. Alternatively, the added disclosures required to make such claims under the NLEA rules may make them less effective or sufficiently costly that producers abandon them. Regardless of the cause, the reduction in claims suggests a reduction in head-to-head competition on nutritional features of food products.

The systematic movement away from comparative claims under the NLEA rules merits further research to better understand the reasons why firms greatly reduced nutrition comparisons in advertising. Research would also be valuable to help determine whether the effects on consumer behavior and on firms' incentives to develop and promote nutritionally preferred foods have been beneficial or harmful. If the earlier claims are misleading to consumers, their elimination should lead to dietary improvements for consumers and to stronger incentives for firms to improve the nutrition profile of foods. If the earlier claims are not misleading but provide useful comparative information to consumers, their loss should slow dietary improvements for consumers and reduce firms' incentives to make food improvements.

In summary, the Commission ordinarily will not challenge comparative nutrient content claims that comply with FDA's regulations, and will carefully scrutinize comparative nutrient content claims that characterize nutrient differences in ways that do not comply with FDA's regulations.

[18] Probit results are consistent with the linear results.

Table 6-3 Regression Results for *Nutrient Content Claims* Across Regulatory Periods[1]

Nutrient	$D_{4/80}$ [FTC ends Part II Food Rule]		$D_{1/83}$ [FTC ends entire Food Rule]		$D_{8/87}$ [FDA health claim proposal]		$D_{2/90}$ [FDA withdraws 1987 proposal/NLEA]		$D_{5/94}$ [FTC Policy Statement/ FDA/NLEA rules effective]	
Fat	.006	(0.55)	-.000	(-0.02)	.065	(6.51)**	.122	(11.40)**	.096	(9.35)**
Saturated Fat	.000	(0.03)	.008	(1.74)*	.025	(5.16)**	.006	(1.07)	-.017	(-3.38)**
Cholesterol	.008	(0.89)	.013	(1.47)	.093	(10.40)**	.036	(3.71)**	-.119	(-12.93)**
Sodium	.015	(1.73)*	.051	(6.34)**	.028	(3.50)**	-.016	(-1.79)*	-.037	(-4.48)**
Fiber	.028	(3.60)**	-.001	(-0.20)	.022	(2.89)**	-.009	(-1.13)	.005	(0.60)
Calcium	-.006	(-1.17)	.019	(4.08)**	.002	(0.46)	.002	(0.48)	.019	(3.85)**
Vitamin/Mineral	-.018	(-2.05)**	.025	(3.07)**	-.011	(-1.32)	.016	(1.75)*	-.000	(-0.00)
Calorie/Diet	.013	(1.15)	.063	(5.91)**	.053	(4.99)**	-.032	(-2.83)**	-.058	(-5.24)**
Significant Coefficients/Total	3/8		5/8		6/8		5/8		6/8	
Number Positive/ Number Significant	2/3		5/5		6/6		3/5		2/6	

Notes. ** indicates significance at the 5 percent level in a 2-tailed test; * at the 10 percent level. t-statistics are in parentheses. Dependent variable equals 1 if ad has claim; 0 otherwise. Linear specification.

[1] Addition of a trend variable does not change qualitative results and trend is significant (positive) only for fat and fiber. Probit estimates give comparable results. The constant terms in the linear estimates are .038 (fat), .008 (sat. fat), .038 (chol.), .021 (sodium), .029 (fiber), .010 (calcium), .070 (vit./min.), .074 (cal./diet). All are significant.

Table 6-4 Regression Results for *Nutrient Comparison Claims* Across Regulatory Periods[1]

Nutrient	$D_{4/80}$ [FTC ends Part II Food Rule]		$D_{1/83}$ [FTC ends entire Food Rule]		$D_{8/87}$ [FDA health claim proposal]		$D_{2/90}$ [FDA withdraws 1987 proposal/NLEA]		$D_{5/94}$ [FTC Policy Statement/ FDA/NLEA rules effective]	
Fat	.004	(0.64)	-.005	(-0.80)	.024	(3.70)**	.041	(5.73)**	.007	(1.01)
Saturated Fat	.001	(0.24)	.008	(2.53)**	.007	(2.25)**	.006	(1.87)*	-.011	(-3.44)**
Cholesterol	.000	(0.09)	.006	(1.61)	.015	(4.26)**	.002	(0.60)	-.026	(-7.22)**
Sodium	.005	(1.07)	.011	(2.55)**	.024	(5.45)**	-.003	(-0.61)	-.023	(-4.95)**
Fiber	.002	(0.64)	.008	(2.33)**	.014	(4.02)**	-.022	(-5.74)**	-.009	(-2.43)**
Calcium	-.003	(-0.97)	.014	(5.12)**	-.004	(-1.33)	-.002	(-0.80)	-.000	(-0.02)
Vitamin/Mineral	.002	(0.65)	-.009	(-2.88)**	-.001	(-0.23)	.016	(4.48)**	-.012	(-3.62)**
Calorie/Diet	.021	(2.80)**	.021	(2.80)**	.018	(2.55)**	-.027	(-3.54)**	-.051	(-6.87)**
Significant Coefficients/Total	1/8		6/8		6/8		5/8		6/8	
Number Positive/ Number Significant	1/1		5/6		6/6		3/5		0/6	

Notes. ** indicates significance at the 5 percent level in a 2-tailed test; * at the 10 percent level. t-statistics are in parentheses. Dependent variable equals 1 if ad has claim; 0 otherwise. Linear specification.

[1] Addition of a trend variable does not change qualitative results and trend is significant (positive) only for fiber and calcium and (negative) for vitamins. Probit estimates give comparable results. The constant terms in the linear estimates are .021 (fat), .000 (sat. fat), .005 (chol.), .001 (sodium), .007 (fiber), .003 (calcium), .012 (vit./min.), .023 (cal./diet). All are significant except for sodium and calcium.

More generally, research on these issues would be useful to regulators and researchers interested in better understanding the effects of triggered disclosures and the role of simple comparative claims in an environment where detailed product information is provided on package labels.

GENERAL NUTRITION CLAIMS AND REGULATION

Background and Method

General nutrition claims, such as "healthy" or "nutritious," are also potentially affected by various regulatory events during the period. Moreover, these general claims can complement or substitute for more specific claims, and thus, could be affected by regulatory changes affecting specific claims. In this section we present a brief summary of the evidence on these general claims across regulatory periods.

Several of the regulatory events could have direct impact on the use of general nutrition claims. The end of Part II of the FTC Food Rule in May 1980 ends the proposal to regulate general health claims broadly. The end of the entire Food Rule in early 1983 terminates proposals to regulate "natural." In both cases we would expect use of general claims to increase following the event. The impact of the health claim events of August 1987 and February 1990 depend on whether general claims complement or substitute for specific health claims; if general claims are usually substitutes for specific claims, the change should be opposite that predicted for specific health claims; and conversely, if they are complements, the effect should be in the same direction. Other events in 1990 include proposals for explicit regulation of general claims and thus would be expected to reduce their use. Finally, the 1994 NLEA rules provide explicit rules for using these claims, and they also impose added restrictions on their use. This would be expected to reduce the use of general claims, unless the removal of the regulatory uncertainty is sufficiently large to dominate this effect.

To examine the relationship between the regulatory events and the use of general claims, we collect claims into five categories of related general claims: *Healthy/Smart/Good for You/Youth*, *Nutritious/Wholesome/Fortified*, *Light/Lean*, *Natural/No Artificial/Real/Pure*, and *Fresh* claims.[19] Table 6-5 presents simple linear regression results in which the presence of a claim from the indicated category is regressed against dummy variables for the major regulatory periods.[20]

Results for General Nutrition Claims: Substitutes for Specific Claims

As predicted, the results in Table 6-5 indicate that the use of general nutrition claims tends to increase in the post-1980 period and in the post-1983 period, as the FTC's Food Rule is ended. Four of the 5 coefficients are significant in each of these periods, and 3 of the 4 in the post-1980 period and 4 of the 4 in the post-1983 period are positive, suggesting significant growth in the use of general nutrition claims from most categories. In the post-1987 period only 2 coefficients are significant and these are both negative (as are two of the

[19] These categories are described in more detail in Chapter 4.
[20] Probit regressions again provide comparable results.

insignificant coefficients). Thus, in the period when the rules governing specific health claims are relaxed, advertisers move away from the use of general nutrition claims to more specific health and nutrient claims, suggesting that general claims and specific claims are substitutes and that specific claims are preferred to the general claims.

Finally in the post-NLEA periods, the use of these claims again changes significantly in four of five equations in both periods. In the post-1990 period the use of general nutrition claims grows (3 of the 4 significant coefficients are positive), evidence that is again consistent with the hypothesis that general and specific claims are substitutes. In the post-1994 period, when added restrictions are imposed by the NLEA rules, the use of general claims drops significantly across four of the groups, and these reductions are all significant and larger than the earlier post-1990 increases. The only exception is for *Nutritious/Wholesome/Fortified* claims, which are not explicitly regulated under the NLEA rules.

Thus, the use of these general nutrition claims drops systematically in the post-NLEA period, though it should be noted that in 1997 more than 55 percent of all food ads still contain general nutrition claims of some type. Systematic movements in the use of general claims relative to specific health claims suggests that general claims are substitutes for specific claims and suggests that, absent other constraints, restrictions on specific claims induce firms to move to more general claims to convey their nutrition message.

Do "Good Foods" Use Health Claims More, Advertise More, Post-NLEA?

Use of Health Claims Across Food Categories

We would expect the changing regulations to affect advertising in some food groups more than others. Health issues are more relevant to some food categories than others and some of the regulatory changes have specific requirement that will restrict health claim use in some food categories.

Table 6-5 Regression Results for *General Nutrition Claims* Across Regulatory Periods[1]

Category	$D_{4/80}$ [FTC ends Part II Food Rule]		$D_{1/83}$ [FTC ends entire Food Rule]		$D_{8/87}$ [FDA health claim proposal]		$D_{2/90}$ [FDA withdraws 1987 proposal/NLEA]		$D_{5/94}$ [FTC Policy Statement/ FDA/NLEA rules effective]	
Healthy/Smart/ Good for You/Young	.018	(1.52)	.042	(3.73)**	.016	(1.36)	.037	(3.02)**	-.066	(-5.57)**
Nutritious/ Wholesome/Fortified	-.029	(-2.57)**	.043	(4.02)**	-.013	(-1.20)	.032	(2.76)**	.017	(1.52)
Light/Lean	.038	(3.10)**	.033	(2.90)**	-.016	(-1.39)	.021	(1.72)*	-.049	(-4.13)**
Natural/No Artificial/ Real/Pure	.072	(4.54)**	.079	(5.26)**	-.026	(-1.71)*	-.053	(-3.25)**	-.062	(-3.97)**
Fresh	.041	(3.45)**	-.010	(-0.88)	-.022	(-1.97)**	-.017	(-1.41)	-.053	(-4.59)**
Significant Coefficients/Total	4/5		4/5		2/5		4/5		4/5	
Number Positive/ Number Significant	3/4		4/4		0/2		3/4		0/4	

Notes. ** indicates significance at the 5 percent level in a 2-tailed test; * at the 10 percent level. t-statistics are in parentheses. Dependent variable equals 1 if ad has claim; 0 otherwise. Linear specification.

[1] Addition of a trend variable does not change qualitative results. Probit estimates give comparable results. The constant terms in the linear estimates are .101 (*Healthy/etc*), .119 (*Nutritious/etc.*), .111 (*Light/Lean*), .028 (*Natural/etc.*), and .144 (*Fresh*). All are significant.

Table 6-6 Regression Results for *Disease and Affiliated Claims* Across Regulatory Periods By Food Category[1]

	Veg/Fruit/ Juice	Cereal/ Bread	Dairy	Poultry/ Fish/Grain	Meat/ Egg	Fats & Oils	Drinks	Sauces/ Dressing/Misc	Desserts/ Snacks
Constant	.000	-.000	.000	.003	.038**	.104**	-.000	.000	-.000
	(0.00)	(-0.00)	(0.00)	(0.65)	(2.66)	(3.31)	(-0.00)	(0.00)	(-0.00)
$D_{5/80}$	-.000	-.000	-.000	.002	-.038*	-.104**	.014**	.011	-.000
(FTC ends Part II FR)	(-0.00)	(-0.00)	(-0.00)	(0.29)	(-1.65)	(-2.11)	(2.82)	(1.22)	(-0.00)
$D_{1/83}$	.027**	.029	.014	-.000	.010	.183**	-.014**	-.004	.001
(FTC ends Food Rule)	(2.26)	(1.22)	(1.18)	(-0.04)	(0.43)	(4.02)	(-2.80)	(-0.47)	(0.62)
$D_{8/87}$	-.015	.165**	-.002	.020**	.103**	.253**	-.000	-.003	.006**
(FDA health claim prop.)	(-1.30)	(7.01)	(-0.15)	(2.98)	(4.81)	(6.28)	(-0.00)	(-0.31)	(2.22)
$D_{2/90}$	.019	-.148**	-.012	-.016**	-.063**	-.278**	-.000	.044**	-.007**
(FDA withdraws 1987 proposal/NLEA)	(1.44)	(-5.68)	(-0.94)	(-2.30)	(-2.79)	(-6.28)	(-0.00)	(4.52)	(-2.53)
$D_{5/94}$	.074**	.044*	.063**	-.009	-.011	-.159**	.000	-.044**	.003
(FTC Policy Statement/ FDA/NLEA rules effective)	(5.05)	(1.83)	(5.17)	(-1.38)	(-0.49)	(-2.97)	(0.00)	(-4.48)	(1.03)
Adj. R^2	.035	.071	.028	.005	.026	.131	.007	.019	.002
Mean Dep. Variable	.024	.055	.016	.007	.040	.171	.002	.013	.001
N	1729	1047	1228	1879	932	720	939	1746	2749

Notes. [1] Product categories are defined in Table 3-2. t-statistics are in parentheses. ** indicates significance at the .05 level; * at the .10 level.

Under the rules developed to implement the NLEA, for instance, health claims are limited to foods that are "best" on the dimensions relevant to the particular health claim, "not bad" on other key dimensions, and "nutritious" in the sense that they provide a minimum level of nutrition on at least one of six specified nutrients (without supplementation).[21] This approach is adopted as part of a strategy to implement the NLEA's goal of educating consumers about healthful dietary practices. By limiting health claims to what might be considered "good foods," it is hoped that producers of these foods will find it more profitable to promote the foods, and as a result, will have greater success in getting consumers to include these foods in their diets in place of less desirable foods. If these presumptions are correct, these NLEA rules should increase the frequency with which sellers of "good foods" use health claims to promote their foods, increase the promotion of these "good foods," reduce the use of health claims by sellers of other foods, and together, these changes could lead to improvements in consumer diets in these relevant food categories.

To examine the advertising part of these hypotheses, Table 6-6 gives the results of simple linear time series regressions that relate the use of *disease and affiliated claims* with the key regulatory events. Regressions are run separately for the nine food categories described in Chapter 3, which together cover all food advertising in our sample.[22]

Between 1980 and 1983, when the FTC focuses the final phase of its rulemaking on health claims, advertisers stop using health claims in the only two categories in which they are not trivial at that time, Meat/Eggs and Fats & Oils (as shown by the equal sized but opposite signed coefficient for the constant term and $D_{5/80}$ in those equations).[23] Once the final Food Rule decision is issued in early 1983, health claims appear in several food categories, though with the exception of Fats & Oils, these claims appeared in fewer than 3 percent of ads during the 1983 to 1987 period. Once the FDA publishes its 1987 proposal for health claims, their use is more widespread and more frequent. The increases are significant in five of the nine categories, in the Cereals/Breads, Poultry/Fish/Grains, Meats/Eggs, Fats & Oils, and Desserts/Snacks categories.

Of interest for assessing the post-NLEA period, however, are the coefficients for the variables $D_{2/90}$ and $D_{5/94}$. As seen in Table 6-6, the use of disease and affiliated claims increases significantly in only one category in the post-1990 period, for the Sauces/Dressings/Misc category, and decreases significantly in five categories, Cereal/Bread,[24] Poultry/Fish/Grains, Meat/Eggs, Fats & Oils, and Desserts/Snacks. After the NLEA rules are finalized and the FTC policy is clarified in May 1994, disease and affiliated claims increase significantly in 3 categories, Fruit/Vegetables/Juice, Cereal/Bread and Dairy, and decrease significantly in two categories, in Fats & Oils and Sauces/Dressings/Misc.

---

[21] See 21 *CFR* 101.14 for general requirements for health claims on labels, or see Ippolito and Mathios (1993) for a summary of the requirements. For example, for a food product to mention sodium's role in hypertension, the product must be "low" in sodium (less than 140 mg per serving); it must contain less than 13 g fat, 4 g saturated fat, 60 mg cholesterol, and 480 mg sodium per serving; and without fortification, it must contain at least 10 percent of the Daily Reference Value for vitamin A, vitamin C, iron, calcium, protein, or fiber.

[22] Because for so many products, no health claims are made during key periods, it is not possible to run directly comparable probit specifications for these estimates. As seen in the next chapter in the case of Fats & Oils, however, the two methods give qualitatively similar results when corrected for periods with no variation.

[23] The few claims in the drink category at this time are somewhat different than the usual positive health claims. One ice tea advertiser at the time makes claims about the absence of artificial sweeteners implicated in cancer risks in laboratory tests. As seen in the subsequent coefficient, these claims are short-lived.

[24] In 1997, the use of health claims increases substantially in the Cereals/Breads category, returning to near 1990 levels, following the FDA authorization of an oat/heart disease claim.

Taking these post-NLEA changes together, the two categories with significant net increases[25] do include foods targeted for increased consumption by public health officials, namely fruits and vegetables and low-fat dairy products, thus providing some evidence of the desired effects from the NLEA. Between 1990 and 1997 the likelihood that Vegetable, Fruit and Juice ads used health claims increases by 9.3 percentage points; Dairy ads increase by 5.1 percentage points.

More sizable effects are found in the food categories that experience decreases following the NLEA. Most significantly, the likelihood that Fat & Oils advertising makes disease or affiliated claims decreases by 43.7 percentage points. The likelihood that Cereal and Bread advertising includes a disease or affiliated claim decreases by 10.4 percentage points; for Meats, Eggs and their substitutes this likelihood falls by 7.4 percentage points; and for Poultry, Fish, and Grains it drops by 2.5 percentage points. The Fats & Oils category, the Meat and Eggs, and the Poultry, Fish, and Grain categories especially include the types of products targeted for reduced health claim use by the NLEA rules. The reduction in these categories is statistically significant, and in the first two categories constitute the largest measurable changes in health claim advertising in the post-NLEA period.

In fact, examination of the underlying data indicates that after 1991 no disease or affiliated claims are included in advertising for any meat entree, meal or individual products, for any poultry, fish, or grain-based entrees, meals, or individual products, or for breads of any kind. Moreover, after 1992 no salad dressings or other sauces, including nonfat dressings, make health claims of any type; and after 1993, no margarine, cooking oil, or other fat product or substitute make such claims. Thus, the evidence indicates that the health implications of choices in all of these food categories appears to have been completely eliminated from advertising under the NLEA rules through 1997.

This evidence suggests that the rules implementing the NLEA produced the desired effect in reducing the focus on health in certain categories that do not meet regulation guidelines, such as Fats & Oils and Meats/Eggs/Mixtures, and to a lesser extent increasing the focus on health in at least one category targeted for increased consumption, Fruits/Vegetables/Juice. In an effort to assess these changes more precisely, we now turn to a more detailed examination of some of the advertising changes in key categories.

Fruit, Vegetable, and Juice Advertising Falls After NLEA; Only Orange Juice Ads Have Greater Health Focus

As described above, after the passage of the NLEA, the percentage of advertising that includes a health claim in the Fruit/Vegetable/Juice category rises significantly, a result that is consistent with the hypothesis that sellers find it more profitable to advertise these products with health claims under the new regulations. However, other data from the category does not support this hypothesis, and in fact, indicates that overall advertising for fruit, vegetables, and juice has fallen under the NLEA.

---

[25] By net increase, we mean products where the post-1990 effects taken together are positive (as indicated by the sum of the $D_{2/90}$ and $D_{4/94}$ coefficients). For instance, despite the positive coefficient for Cereal/Bread in 1994, the magnitude is not large enough to overcome the negative coefficient in 1990. The same is true for the Sauces/Dressing/Misc. category.

The first type of evidence is the number of different types of fruits and vegetables that make health claims in advertising. Orange juice producers are the most frequent users of health claims throughout the years of the study, but in the years after the NLEA rules, health claims in the category are used almost exclusively by orange juice producers.[26] Prior to the NLEA rules, a few other fruit or vegetable producers make heart or cancer claims in our sample, including grapefruit juice producers, West Coast pear producers, California lima bean producers, and the California Prune Board. Prior to passage of the NLEA in 1990, 61 percent of advertising with a disease or affiliated claim in the Fruit/Vegetable/Juice category is for orange juice; after the NLEA rules are effective in May 1993, 95 percent of ads with these claims in the category are orange juice claims. Thus, the evidence is not consistent with the hypothesis that the post-NLEA rules increase the number of different types of fruit and vegetable producers using health claims in their advertising; in fact, the opposite is true.

Second, as illustrated in Figure 6-1, the amount of advertising in the Fruit/Vegetable/Juice category falls significantly after the NLEA. Between 1977 and 1990, our sample includes between 78 and 120 advertisements per year in the category, and while variation exists from year to year, the data do not indicate any discernable trend. After 1990 the number of advertisements in the category falls significantly and substantially, stabilizing at less than 50 advertisements per year on average. The number of advertisements in the category in the post-NLEA period falls by approximately 50 percent compared to the pre-NLEA period.

Figure 6-1 Number of Advertisements for the Fruit/Vegetable/Juice Category[1]

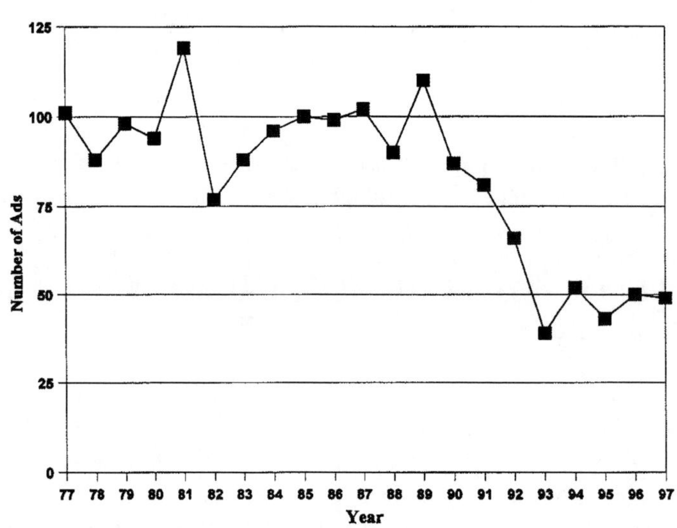

[26] After 1990, with only two exceptions, all advertisements in the category that have health claims are orange juice ads. One advertisement for Campbell's V8 juice includes a cancer and heart claim in 1996, and one advertisement for the California Dry Beans Association has a heart claim.

Figure 6-2 Fruit/Vegetable/Juice Ads As a Percentage of All Food Ads

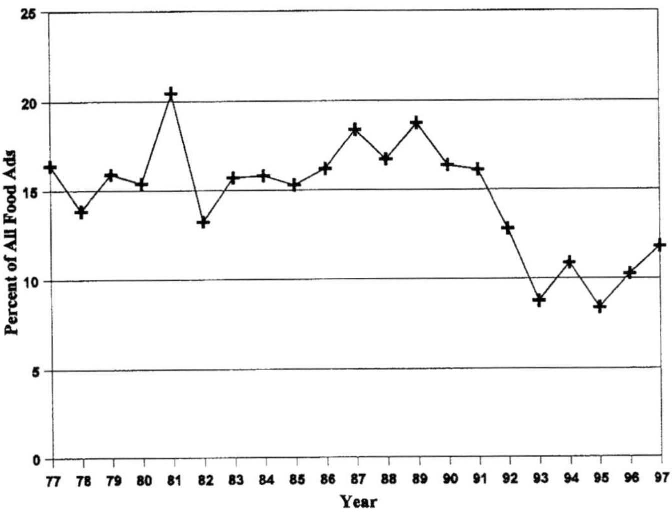

Notes. [1] The *Fruit/Vegetable/Juice* category includes all fruit, vegetable or juice products, as described in Table 3-2.

Recall from Chapter 3 that food advertising in general exhibits a downward trend over the years of the study. Nonetheless, the changes in the Fruit/Vegetable/Juice category are substantially more pronounced that those for food advertising in general. Figure 6-2 illustrates the percentage of food advertising in the Fruit/Vegetable/Juice category in each year. The data illustrate the same pattern as the advertising count data: Fruit/Vegetable/Juice advertising is approximately 17 percent of all food advertising through 1990 and then falls to approximately 10 percent of food advertising after 1990, a 35 percent reduction.[27]

Thus, taken together the evidence for the Fruit/Vegetable/Juice category indicates a significant reduction in advertising for this category in the post-NLEA period; the number of ads in the category drops by half and the percentage of food advertising in the category drops by about one-third. Producers that continue to advertise are more likely to use health claims, though these are almost exclusively orange juice producers in the post-NLEA period.

Advertising Falls or Remains Stable in All Food Categories in Post-NLEA Period

The hypothesis that added regulatory restrictions on health claims in the post-NLEA period make it easier and more profitable for firms selling "good" foods to advertise, leading to more advertising by these types of foods, is rejected more generally by the data examined here. For instance, Table 6-7 presents simple linear regressions relating the number of

[27] We examined detailed data on the particular firms advertising in this category over time. We did not find any large advertisers or particular classes of fruit or vegetable producers who systematically stopped advertising between 1990 and 1993 that could account for the data shift. In particular, we did not find the drop concentrated among the marketing order advertisers, which we checked in light of litigation against marketing orders during this period. See Cal-Almond v. USDA (1993) and the Supreme Court opinion in Glickman v. Wileman (1997), or the summary in Crespi (2001).

advertisements per month with the key regulatory events. Regressions are again run separately for the nine food categories and the results are shown in Table 6-7.[28]

Prior to 1987 coefficients for the regulatory dummies are generally not significant, with the exception the 1980 dummy for the Poultry/Fish/Grain category. With this one exception, these results indicate that the number of advertisements does not change significantly in the pre-1987 environment. Between 1987 and 1990, when the FDA labeling rules are relaxed and health claim advertising is at its peak, the number of advertisements in the listed food categories shows no significant increases or decreases, except in the Desserts/Snacks category, where the number falls by 34 percent.

Finally, in the post-NLEA period, 8 of the 9 coefficients on the $D_{2/90}$ variable are negative (the exception is an insignificant positive for Desserts/Snacks). Three of these decreases are significant, the Vegetables/Fruit/Juice category, as discussed above, the Cereal/Bread category, and the Fats & Oils categories. For the NLEA final event in May 1994, 6 of 9 coefficients are negative, and 2 are significant, Fats & Oils and Fruit/Vegetables/Juice. Three of the coefficients are positive, but they are all of the same magnitude and opposite in sign from their corresponding 1990 coefficients, indicating no net change for all three in the post-NLEA period.

Thus, at this level of aggregation, there is no evidence of increased advertising in "good" food categories in the post-NLEA period, but some evidence of reduced advertising across certain food categories. These reductions are significant in one category targeted for reduced consumption (Fats & Oils), as well as one targeted for increased consumption (Fruit/Vegetables).

Health Claims for Desserts and Snacks Have Not Changed Under the NLEA; Evidence Indicates Health Claims Are Not Used for These Foods

A number of provisions in the NLEA rules are motivated by a concern that producers of empty or otherwise nutritionally deficient foods would use health claims in marketing their products. For instance, the requirement that foods making health claims on labels must have at least 10 percent of the Daily Reference Value of vitamin A, vitamin C, iron, calcium, protein, or fiber[29] is commonly called the "jelly bean rule." This reflects the fact that without the nutritional requirement, an advertiser of jelly beans (or other sugar-based products) could make (say) a heart disease claim under NLEA rules, because the product is low in fat and saturated fat and contains no cholesterol. In the late 1980s, when the health claim debate was most vigorous, the prospect of oat bran potato chips with heart claims, or fiber enriched donuts with cancer claims, were regularly invoked as part of the rationale for stricter regulation.

---

[28] Other specifications which allowed for an overall time trend or control for the cost of magazine advertising (as described in Chapter 3) do not change the findings reported here. None of the food categories have a significant time trend in those specifications.

[29] See 21 *CFR* 101.14.

Table 6-7 Regression Results for Number of Ads per Month Across Regulatory Periods By Food Category[1]

	Veg/Fruit/ Juice	Cereal/ Bread	Dairy	Poultry/ Fish/Grain	Meat/ Egg	Fats & Oils	Drinks	Sauces/ Dressing/Misc	Desserts/ Snacks
Constant	32.60**	20.30**	17.60**	34.20**	18.60**	12.50**	21.10**	31.40**	49.90**
	(11.86)	(11.51)	(7.82)	(12.01)	(10.46)	(8.75)	(11.72)	(10.11)	(13.21)
$D_{5/80}$	-1.23	-4.18	4.65	-9.58**	-4.23	-1.75	-3.35	2.10	8.23
(FTC ends Part II FR)	(-0.30)	(-1.58)	(1.38)	(-2.24)	(-1.58)	(-0.82)	(-1.24)	(0.45)	(1.45)
$D_{1/83}$	0.84	0.95	-1.96	5.23	0.41	2.89	-3.32	-3.50	-6.98
(FTC ends Food Rule)	(0.22)	(0.38)	(-0.62)	(1.31)	(0.16)	(1.44)	(-1.32)	(-0.80)	(-1.32)
$D_{8/87}$	0.66	0.30	0.09	0.52	1.84	2.11	-1.18	-1.25	-17.64**
(FDA health claim prop.)	(0.17)	(0.12)	(0.03)	(0.13)	(0.74)	(1.05)	(-0.47)	(-0.29)	(-3.33)
$D_{2/90}$	-10.79**	-4.88*	-3.46	-1.79	-3.21	-5.25**	-1.00	-4.67	3.33
(FDA withdraws 1987 proposal/NLEA)	(-2.72)	(-1.92)	(-1.07)	(-0.44)	(-1.25)	(-2.55)	(-0.38)	(-1.04)	(0.61)
$D_{5/94}$	-6.36*	4.50*	3.45	1.96	-1.59	-4.50**	-0.34	-3.63	-4.20
(FTC Policy Statement/ FDA/NLEA rules effective)	(-1.75)	(1.93)	(1.16)	(0.52)	(-0.68)	(-2.39)	(-0.14)	(-0.88)	(-0.84)
Adj. R^2	.341	.092	-.017	.006	.064	.269	.198	.110	.354
Mean Dependent Variable	27.4	16.6	19.5	29.8	14.8	11.4	14.9	27.7	43.6

Notes. [1] Product categories are defined in Table 3-2. t-statistics are in parentheses. ** indicates significance at the .05 level; * at the .10 level.

The extent to which desserts or snack foods do indeed make health claims can be examined with the data in our sample. To this end we focus on two of our food categories, Drinks, which includes all carbonated soft drinks and all fruit-flavored beverages (but not juice), along with other beverages such as coffee, tea and water, and Desserts/Snacks, which includes desserts, sweets, donuts, danish and other sweet breads, salty snacks, such as potato chips and related items. Most so-called "junk foods" are included in these two categories, along with many foods that have positive nutritional value.

Table 6-8 presents the percentage of advertisements in each year with a disease or affiliated claim in the two categories. First, note that in every year but three for the Dessert/Snack category, and in every year but one for the Drink category, no advertisements contained disease or affiliated claims. Thus, throughout the regulatory periods, disease and affiliated claims are not a significant phenomenon in these categories.

Table 6-8 Percentage of Advertisements with *Disease or Affiliated Claims* for Desserts/Snacks and Drink Categories

Year	Desserts/Snacks	Drinks
1977	0.0	0.0
1978	0.0	0.0
1979	0.0	0.0
1980	0.0	3.6[1]
1981	0.0	0.0
1982	0.0	0.0
1983	0.0	0.0
1984	0.0	0.0
1985	0.5[2]	0.0
1986	0.0	0.0
1987	0.0	0.0
1988	0.0	0.0
1989	2.2[3]	0.0
1990	0.0	0.0
1991	0.0	0.0
1992	0.0	0.0
1993	0.0	0.0
1994	0.0	0.0
1995	0.0	0.0
1996	0.0	0.0
1997	1.2[4]	0.0

Notes. [1] These advertisements for a low calorie lemonade focused on the absence of an artificial sweetener, with a claim that the sweetener "had been determined to cause cancer in laboratory animals."

[2] An ad for peanut butter included the tag line "good nutrition straight from the heart."

[3] Ads for a low fat, low cholesterol oat bran muffin mix with the claim "as part of a low fat, low cholesterol diet, can help reduce cholesterol" and as much fiber "as a bowl of fruit-bran cereal."

[4] Ads for low fat rice cakes promoted a national walk to raise funds for breast cancer research.

Moreover, the occasional exceptions are either likely to be allowed under the NLEA or are marginal claims picked up by our coding system, which may or may not have been seen as health claims by consumers. In the latter category are the 1985 advertisements for a peanut butter with the tag line "good nutrition straight from the heart" and the 1997 advertisement for a rice cake product providing publicity for a walk to raise funds for breast cancer research. Recall that our coders are not allowed to judge the intent of any claim in the context of an ad, but are required simply to code the presence of any disease-related words in the advertisements, hence the heart and cancer claims in these cases. The 1989 advertisements are for a low fat, oat bran muffin mix with the claims "as part of a low fat, low cholesterol diet, can help to reduce cholesterol" and as much fiber "as a bowl of fruit-bran cereal." It is impossible to judge from the advertisement whether the product would meet the current requirements for an oat bran-heart disease claim but certainly that is possible. The 1980 advertisements in the Drink category are ads for a low calorie lemonade drink highlighting the absence of an artificial sweetener, which "has been determined to cause cancer in laboratory animals." In the late 1970s and early 1980s, concern about artificial sweeteners led to claims of this type in a few product categories.

Certainly, these data do not support the hypothesis that absent strict regulatory restraints, health claims would be widely used by producers of nutritionally vacuous or significantly deficient products. In part, this lack of claims may reflect advertisers' concerns about normal advertising enforcement against deceptive or misleading claims. Alternatively, such claims may not be effective with consumers who presumably might be skeptical of claimed health benefits of oat bran potato chips or the like. Whatever the cause, the evidence indicates that both before and after the NLEA, health claims have not been a significant phenomenon in the Dessert/Snack or Drinks categories.

Finally, Figures 6-3 and 6-4 present data on the amount of advertising in these two categories. The Desserts/Snacks/Sweet Bread category is a large category with considerable advertising, averaging more than 50 advertisements per month in the early years of the sample. As reflected in the regressions reported in Table 6-8 and the data in

Figure 6-3 Number of Advertisements for the Dessert/Snack/Sweet Bread Category[1]

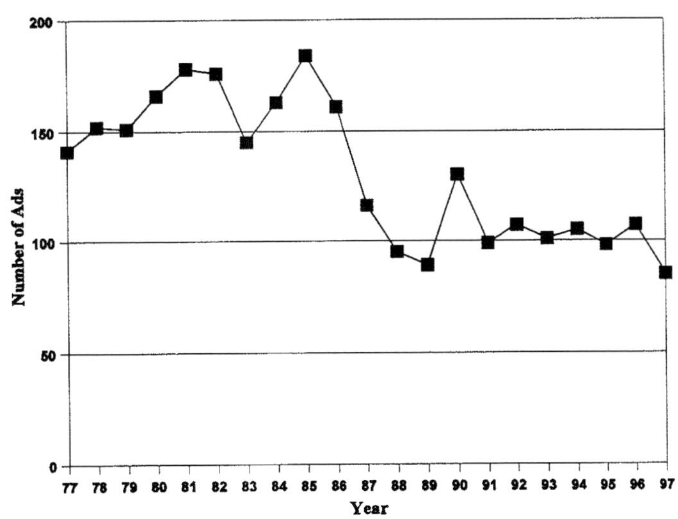

Figure 6-4 Number of Advertisements for the Soft Drink Category[2]

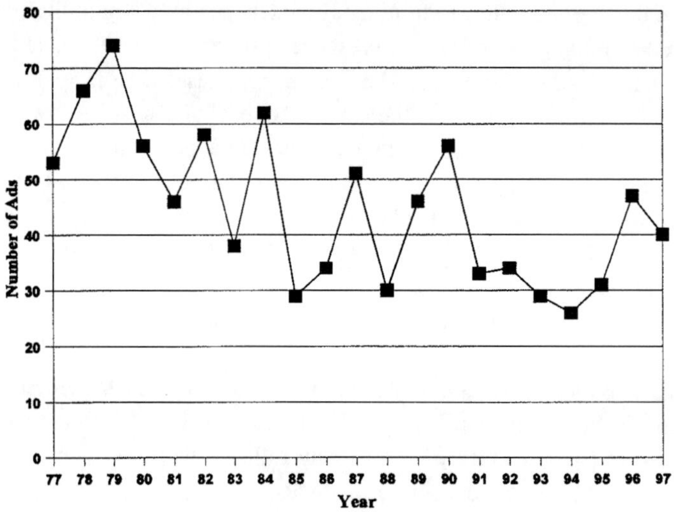

Notes. [1] The *Dessert/Snack/Sweet Bread* category includes all dessert and snack items, including candy, sweets, nuts, muffins, chips, and related items, as described in Table 3-2.

[2] The *Soft Drink* category includes all coffee, tea, soda, fruit flavored drinks, *etc*. It does not include milk, juice, or alcoholic beverages.

Figure 6-3, however, the amount of advertising in the Desserts category falls significantly in the late 1980s, the period of greatest health claim advertising, and this reduced level of advertising does not change in the post-NLEA period. The number of advertisements in the category falls by approximately one-third during the late 1980s. For Drinks, there is a general slow decline in the number of advertisements over the whole period from approximately 21 advertisements per month in the late 1970s to approximately 12 advertisements per month in the post-NLEA period. The decline does not appear to be particularly associated with any of the regulatory periods, as seen in the regression in Table 6-8.

Thus, both the Desserts/Snacks category and the Drinks category exhibit less advertising over time, but these declines do not seem to be associated with the NLEA or its implementing regulations. In fact, the substantial reduction in the Desserts/Snacks category coincides with the period of greatest health claim activity in the late 1980s.

CONCLUDING REMARKS

Overall, the results here indicate that the content of food advertising varies considerably with changes in regulation and enforcement. Advertising claims about the health implications of diet have experienced the most dramatic movement over the years of this study, as the rules shifted from prohibitions in the early years, to normal deception standards in the late 1980s, and then back to a more qualified acceptance under the NLEA rules of the later 1990s.

Nutrient content claims also vary with regulatory changes, even when the regulatory changes do not directly affect the use of such claims, as when the rules for health claims change but not those for nutrient content claims. The reason for these results deserves further study. Health claims deal directly with the consequences of particular characteristics, as in the case of heart disease as a consequence of saturated fat consumption. It may be that without that explicit reminder of the reason to care about a particular nutrient, consumers are not as responsive to simple nutrient claims. Alternatively, the consequence claims may provide direct information to consumers and that without it, the nutrient claims are not meaningful.

Under the NLEA rules advertisers have shifted away from most other nutrients to focus more exclusively on total fat claims. Also, advertisers have shifted away from comparative claims on all dimensions, again with the exception of total fat. Which features of the rules are associated with the movements and whether these changes are desirable for consumers are ripe areas for further research.

We do see evidence of substitution between general nutrition claims, such as *wholesome*, and more specific nutrient claims depending on the regulatory environment. When specific claims are more difficult or more costly, advertisers shift to more general nutrition claims, suggesting that general nutrition claims are often a second best choice to advertisers.

Finally, the evidence here does not support the hypothesis that one of the benefits of tightening the requirement for health claims to only the best foods would lead to "good" foods, such as fruits and vegetables, increasing their advertising and their focus on health in advertising. Post-NLEA advertising for fruits and vegetables falls and only orange juice producers make health claims in our data.

Chapter 7

ECONOMICS OF ADVERTISING: ISSUES AND EVIDENCE

INTRODUCTION

Advertising is a major feature of consumer good markets. When George Stigler (1961) wrote his now-classic paper on the economics of information, economists viewed advertising primarily as a barrier to entry. At the time it was considered revolutionary to think that economic forces applied to information and that market institutions like advertising might be a response to the market's need for information.[1] Since then, the economic view of advertising has changed fundamentally.[2]

As in the Stigler paper, part of the literature addresses the direct information function of advertising, focusing on its potential to inform consumers about product characteristics. For instance, Butters (1977) theoretically examines the informational role of price advertising in shaping market equilibrium prices, Ehrlich and Fisher (1982) stress advertising's role in economizing on shopping time, and Grossman (1981) highlights firms' incentives to provide information created by the ability to credibly advertise product characteristics.

Other papers, typified by Nelson (1970, 1974), Klein and Leffler (1981), Kihlstrom and Riordan (1984), Ippolito (1990) and others, focus on the indirect ways in which advertising can signal product quality. Firms' large public advertising expenditures can sometimes credibly signal a commitment to produce goods with promised characteristics, even when consumers cannot judge quality at purchase.

In this signaling view, the content of advertising may not be important. Information is provided by the expenditure itself. In fact, a number of the signaling papers, as well as some more general advertising papers, begin with observations that it is "obvious" that many ads provide essentially no information.[3]

Empirical studies in the literature have attempted to assess whether prices (say) are lower or higher in markets where advertising is allowed. This work is typified by the Benham (1972) study of prices for optometric services, which finds that prices are lower in states that

[1] In Stigler's words, "advertising ... is treated with a hostility normally reserved for tariffs or monopolists." A notable exception at the time is seen in Telser (1962), where advertising is found to be a means of competition.

[2] For a recent review of the economics of advertising literature, see Bagwell (2001).

[3] See, for instance, Nelson (1974, p. 732), Kihlstrom and Riordan (1984, p. 427) or Becker and Murphy (1993, p. 943).

allow advertising compared to those that do not, and the Bond *et al.* (1980) study of the same market, which finds that prices are lower and the quality of services is comparable in states that allow advertising.[4] Similar in spirit are studies of prescription drugs by Cady (1976), retail gasoline by Maurizi (1972), grocery prices during a newspaper strike by Glazer (1981), and liquor prices in Rhode Island before and after a legal decision removing an advertising ban by Milyo and Waldfogel (1999).

But these tests, and others like them, are all indirect tests of advertising's roles. This empirical work does not directly address the question of whether advertising contains informative claims about products, or works through more indirect methods as typified by the signaling literature. In large part, the difficulty of obtaining data on the types of claims actually made in advertising inhibits more direct tests.[5]

In this chapter, we provide several types of evidence on the hypothesis that advertising plays a direct information role in markets and that it is shaped by economic forces. Using our detailed data on the types of claims made in magazine food advertising, we show that a large portion of this advertising has claims about specific product characteristics. As we saw in the last chapter and will see further in the evidence below, producers change the content of their advertising systematically in response to what is or is not allowed under existing regulations, evidence suggesting that producers believe that the content of their advertising is important. Finally, we present several types of evidence consistent with the unfolding theory of advertising, that is, evidence that competitive forces push producers to reveal more about their products and to bring that information to potential customers when free to do so easily.

INFORMATION IN ADVERTISING: DIRECT EVIDENCE

Background

As consumers search among available products in making purchase decisions, firms have an incentive to try to draw consumers to consider their products, especially consumers who would ultimately become regular customers. By providing product information through spending on advertising, firms can attract consumers who find these advertised characteristics desirable. Firms with products that deliver the advertised characteristics are more likely to get the repeat business necessary to make the advertising worthwhile. This simple mechanism is the fundamental force underlying the information theory of advertising.

The range of information that could be conveyed from sellers to potential customers is extensive. Sellers can inform, or remind, consumers of the existence of the product. Sellers can convey specific attributes of the product (*e.g.,* for foods, nutritional features, taste, varieties, price, ease of use, *etc.*). They can also convey how the product might be used or why it might be valuable to consumers (*e.g.,* for foods, by providing recipes or suggestions for use, or by informing consumers about the health benefits of the product). Firms can also use advertising to identify where the product is sold and other characteristics of the seller.

[4] But see Parker (1995) who questions the quality finding.

[5] To our knowledge no public datasets have systematic data on the claims made in advertising. Several papers in the marketing literature do contain content information collected directly by the authors to address topics under study by them. These include Pappalardo and Ringold (2000), Laband (1989), Dowling (1980), and Resnik and Stern (1977).

Obviously, some media are more suited to providing some types of information and not other types, but with a mix of media, a firm can lower consumers' cost of acquiring a great range of information about its products.

Specific Nutrition Claim Measures

The data collected for this study includes several categories of claims about specific product characteristics that would be relevant to consumers in choosing food products. To consider nutritional characteristics from this perspective, we create several indices for claims about specific nutritional characteristics of foods. As shown in Table 7-1, the first index, *Lipids*, focuses on the 5 main lipid (fat) dimensions of foods and counts the number of distinct lipids featured in the ad. The second index, *Main nutrients*, adds 6 other major nutritional features of foods to the lipids (sodium, fiber, calcium, carbohydrates, protein, and calories), as well as other vitamins and minerals treated as a group. Finally, the third index, *Main nutrients & individual vitamins and minerals*, is comparable to the second index except that individual vitamins and minerals are considered in more detail. Each index reflects the number of distinct dimensions with specific claims in the advertisement.

In some of the analysis, we also use indicator variables for each category of claims that indicate the presence of claims for *any* of the covered nutrient dimensions. Thus, for instance, the *Main nutrients indicator* variable equals 1 if the ad has a claim for any of the main nutrients, and is zero otherwise. The other indicator variables for *Lipids* and *Main nutrients & individual vitamins and minerals* are comparable.

Recall also from Chapter 4 that our coding system also collected a summary measure of whether an advertisement contains any *specific* nutrition-related claim. This measure, labeled *Specific nutrient, health, fat, oil, or calorie claims*, indicates all ads with a claim coded in the specific nutrition-related claim part of our coding scheme. Thus, this index reflects main nutrient claims, as described above, as well as specific health claims, other specific nutrition-related claims, such as *lactose free* or *contains wheat germ*, and fat and oil claims, such as *made with canola oil*.[6] Advertisements that contain only general nutrient claims, as typified by the terms *nutritious* or *healthy*, are not included in this measure.

[6] See the text surrounding Figure 4-22 for a more detailed discussion.

Table 7-1 Indices for Lipid and Nutrient Claims

Index[1] / Indicator[2]	Nutrients Included in Index
Lipids	Total Fat
(0 to 5) / (0 or 1)	Saturated Fat
	Monounsaturated Fat
	Polyunsaturated Fat
	Cholesterol
Main Nutrients	Lipid Dimensions (5) plus
(0 to 12) / (0 or 1)	Sodium
	Fiber
	Calcium
	Any other vitamin or mineral
	Carbohydrates
	Protein
	Calorie
Main Nutrients & Individual	Main Nutrients (11)[3] plus
Vitamins or Minerals	Vitamin B
(0 to 21) / (0 or 1)	Vitamin C
	Vitamin E
	Beta Carotene
	Potassium
	Antioxidants
	Iron
	Folic Acid
	General/Multiple Vitamin
	Other Specific Vitamin

Notes. [1] Each index can take a value from 0 to the specified maximum value depending on the number of listed dimensions for which claims are made in the ad.

[2] Each indicator variable equals 1 if the ad contains a claim for *any* of the listed nutrients in the category, 0 otherwise.

[3] The vitamin and mineral dimension listed in *Main Nutrients* is replaced by the vitamin and mineral dimensions listed here, and thus, represents a more detailed index.

Specific Nutrition Claims Have Become a Major Feature of Food Advertising

Figure 7-1 presents evidence on the percentage of advertisements in each year that contain a *Main Nutrient Claim* or a claim from the broader category of *Specific nutrient, health, fat, oil, or calorie claims*. Both measures demonstrate a substantial increase in the percentage of advertisements that include specific nutrition-related claims during the first half of our period. Since the late 1980s, however, the percentage of ads with nutrition claims has stabilized; approximately 40-50 percent of ads include claims about main nutrients, and approximately 50-60 percent include claims from the somewhat broader class of specific nutrition-related claims.

Using either measure, the evidence indicates that despite a changing regulatory environment, a substantial portion of magazine food advertising contains specific claims about nutritional features of food products. Since the late 1980s, approximately half of all food ads have at least one specific nutrition-related claim.

Other Specific Informative Claims Are Also Common in Food Advertising

Recall also from Chapter 3 that our coders recorded the presence of a number of other types of claims that appear in food advertising. Of particular interest for the current discussion are specific claims about the availability of different varieties of the product, suggestions for using the product, claims highlighting convenience or ease of preparation, claims indicating that the product is new or has been improved, and claims about the taste, aroma, or texture of the product. In each of these cases, producers are attempting to attract consumers interested in these specific features of the product.

Table 7-2 summarizes evidence on the 2 nutrition-related categories illustrated in Figure 7-1, as well as the other specific product claims mentioned above. The table gives the overall percentage of advertisements in the sample that have a claim from the category, as well as the minimum and maximum percentages of ads per year with claims for the years from 1977 to 1997. While some variation exists from year to year, the variation is greatest in the nutrition-related categories, reflecting the general growth in the use of nutrition claims that occurred in the first 10 years of the sample.

Figure 7-1 Percentage of Advertisements with at Least One Specific Nutrition Claim by Type[1]

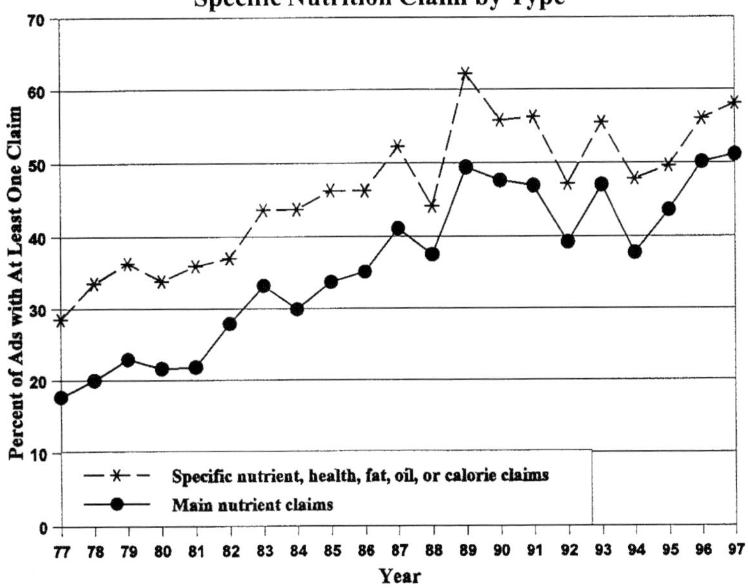

Notes. [1] *Main nutrient claims* are specific claims about the nutrients listed in Table 7-1. *Specific nutrient, health, fat, oil, or calorie claims* includes all claims in the specific nutrition-related claims portion of our coding scheme, including all ads with main nutrient claims, as well as specific health claims, other specific nutrition-related claims, such as *lactose free*, and other fat or oil claims, such as *made with canola oil.*

As shown in Table 7-2, approximately 40 percent of ads include specific suggestions for using the product, often by providing recipes that use the food. More than half of the advertisements include information about the existence of different varieties of the product. Approximately one-third of the ads make a claim about the product's convenience for some use. Approximately 20 percent of the ads highlight that the product is new or has been improved. Finally, approximately 80 percent of advertisements make a claim about the taste, texture, or aroma of the food.

Taken together, this evidence illustrates that a great many advertisements make specific claims about the advertised product. Assuming that the nutrition label is credible, consumers can verify nutrition claims with nutrition information on the label. Consumers can also experiment with suggested uses of the food, determine whether ease of use and convenience claims are valid, try new or improved products, and judge taste or other sensory characteristics. Thus, as would be expected, most of these claims involve search or experience characteristics[7] that consumers can judge before or after purchase.

Table 7-2 Percentage of Food Ads with Specific Information Claims by Type[1]

Type of Claim	Percentage of Ads	
	Overall Sample	Min / Max [2] Annual Percent
Nutrition-Related Categories		
Main nutrient claims [3]	35.0	18 / 50
Specific nutrient, health, fat, oil, or calorie claims [4]	45.5	28 / 62
Other Categories of Specific Product Claims [5]		
Variety information[6]	51.8	44 / 64
Suggestions for use[7]	40.6	32 / 50
Convenient/quick/easy claims[8]	32.4	25 / 41
New/introducing/improved claims[9]	20.7	13 / 27
Taste/aroma/texture claims[10]	79.6	65 / 87

Notes. [1] Data from magazine advertising sample.
[2] Range of annual percentages for years from 1977 to 1997.
[3] Includes all ads with a *main nutrient claim* as defined in Table 7-1.
[4] Includes all ads with a specific nutrient-related claim, as defined in the text surrounding Figure 4-22.
[5] These categories of claims are described in more detail in Chapter 3.
[6] *Variety claims* category includes all explicit claims about the varieties available, including package sizes, flavors, *etc.*
[7] *Suggestions for use* category includes all explicit suggestions for using the product, including recipes.
[8] *Convenient/quick/easy claims* includes all claims about ease of use.
[9] *New/introducing/improved claims* includes all claims about new or improved product.
[10] *Taste/aroma/texture claims* include all claims about the taste, aroma, or texture of the product.

[7] As defined in Nelson (1970).

ADVERTISING AND UNFOLDING: DOES COMPETITION LEAD TO INFORMATION ON MORE NUTRITION DIMENSIONS?

Theory and Method

One of the economic issues in advertising is the potential bias in the types of information provided by advertisers; advertisers have an incentive to tell potential customers what is good about their product but not what is bad about the product. This issue is of particular concern in multi-attribute products, such as foods, where claims about the desirable features could draw attention away from less desirable and unrevealed characteristics. Despite the inherent bias at the individual firm level, economic theory suggests that in many cases competition among producers can substantially reduce or eliminate this bias in the information provided by the market as a whole (Grossman 1981).

For instance, this theory would predict that if firms advertise the no-cholesterol benefits of their products and are gaining sales by omitting information on other dimensions, such as saturated fat, competing firms with low cholesterol, low saturated fat products have the incentive to highlight these facts. This "unfolding" theory suggests that despite firms' initial reluctance to highlight "bad" nutritional characteristics, competition will often induce all but the worst products to disclose key features of available products. As long as consumers are skeptical of firms that do not reveal key information,[8] the market would induce the firm to fill in important missing information, both cholesterol and saturated fat in our example.

The concern about partial information and selective highlighting of particular nutrient dimensions underlies some of the changes implemented in the NLEA rules for food labeling. For instance, under the NLEA, if producers of certain products make any nutrient claims on their labels, they are also required to disclose undesirable characteristics. Thus, for instance, if a claim is made about the low saturated fat content of a high fat product, a disclosure would have to be made referring consumers to the fat content on the nutrition label. The rule is designed to induce more complete nutrition information into the market. The presence of these labeling rules creates greater pressure on firms to consider these issues in advertising. Of course, triggered disclosures of this type also reduce the incentive to make the original nutrient claims at all, because the claims are now more costly in space and complexity. This could have the unintended consequence of reducing competition on nutrition overall.

The data in our sample allows us to examine the unfolding hypothesis in several ways. First, we examine the number of different nutrient dimensions about which claims are made in advertising over time. In particular, we examine whether the number of nutrient dimensions included in advertising increased during the late 1980s, when the regulatory constraints are relaxed and disease and affiliated claims are at their peak, presumably increasing competition on the nutritional characteristics of foods. We also examine whether the number of nutrients mentioned in advertising increases still further in the post-NLEA period when triggered disclosures are specified by regulation in an effort to fill in key nutrition information in claims.

The nutrient measures used in this analysis are those listed in Table 7-1, namely the *Lipids index*, the *Main nutrients index*, and the *Main nutrients & individual vitamins and*

[8] Consumer survey evidence suggests that most consumers view advertising as a selling message and bring considerable skepticism to interpreting advertising claims. See, for instance, Calfee and Ringold (1994).

minerals index. Each index can take on an integer value from zero to the maximum number of nutrients in the category (as listed in Table 7-1) and indicates the number of different dimensions for which claims are made in the advertisement. Figure 7-2 illustrates the mean number of nutrients for which claims are made for each of the three indices for each year of our sample.

Lipid Dimensions

The mean number of lipid dimensions in ads increases only slightly between 1977 and 1987, but then it rises substantially from .13 dimensions per advertisement in 1987 to .57 dimensions per advertisement in 1991, more than 4 times higher. This mean falls by about 20 percent between 1991 and 1997, to .47 dimensions per advertisement.

The mean number of lipids per advertisement is the product of two factors, the percentage of ads that have *any* lipid claims and the average number of lipid dimensions in ads that have a lipid claim. Figure 7-3 illustrates the second of these two factors, the conditional mean, that is, the average annual number of lipid dimensions in advertisements for ads that have at least one lipid claim. Obviously this mean is always greater than one, since only ads with at least one lipid dimension are included in the mean. As shown by the bottom line in the figure, the number of lipid dimensions in ads grows in the mid-1980s from 1.31 dimensions per ad in 1983 to 1.65 dimensions per ad in 1991, but falls back to 1.26 dimensions per ad by 1997. Thus, this evidence makes it clear that for advertisements that make lipid claims, the number of lipid dimensions included in the ads rises through the passage of the NLEA, but then falls substantially through the 1990s returning to the level of the early 1980s.

Figure 7-2 Mean Number of Nutrients per Advertisement by Category[1]

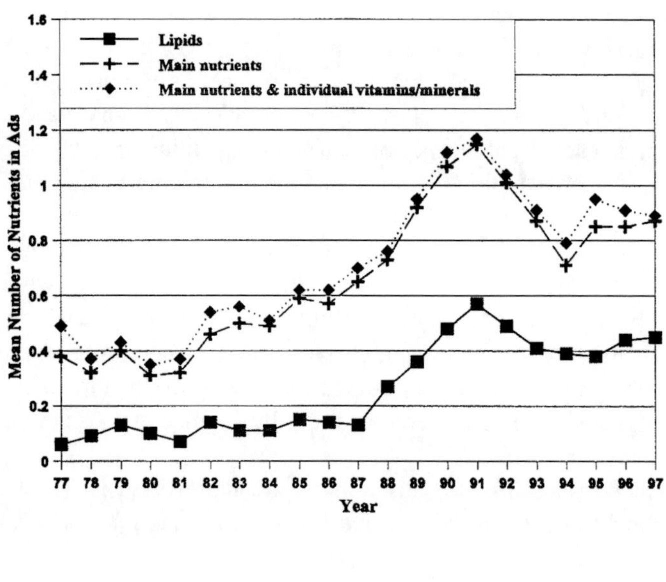

Figure 7-3 Mean Number of Nutrients per Advertisement for Ads with at Least One Nutrient Claim by Category[1]

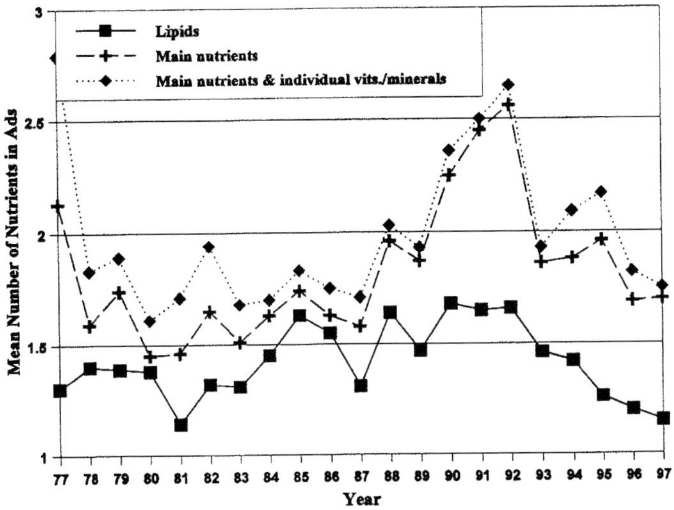

Notes. [1]Nutrient categories are defined in Table 7-1.

Figure 7-4 illustrates the first factor of the overall mean, the percentage of advertisements that contain any lipid claims. The percentage of ads with a lipid claim changes little until after 1987, when approximately 10 percent of advertisements have a lipid claim. By 1991, the percentage of ads with a lipid claim increases substantially to 34.4 percent of ads, and after a small reduction in the early 1990s, the number increases further to 39.5 percent of all ads in 1997.

Thus, the evidence on lipids indicates that there is a rapid increase in the competitive focus on lipids in the late 1980s, reflected in both the percent of ads that make a lipid claim and the number of dimensions included in the ads that make a claim. After the NLEA, lipids remain a focus of competition, but only for a single lipid dimension at a time, typically total fat.

The extent of this reduction in multidimensional competition in lipids is illustrated in Table 7-3, which gives the percentage of ads per year that have claims for one or more than one lipid dimension for the years 1977, the first year of our sample, 1983, before the mid-1980 changes, 1991, at the end of the late-1980 more relaxed regulatory period, and 1997, after the NLEA rules are fully implemented. These data illustrate the extent to which producers are induced to focus on more than one lipid dimension by the competition of the late 1980s and the extent to which this has faded in the post-NLEA period. In 1991, 20.1 percent of advertising had claims for more than one lipid dimension, typically saturated fat and cholesterol; by 1997, only 5 percent of ads had claims about more than one lipid dimension.

Figure 7-4 Percentage of Advertisements with at Least One Nutrient Claim by Category[1]

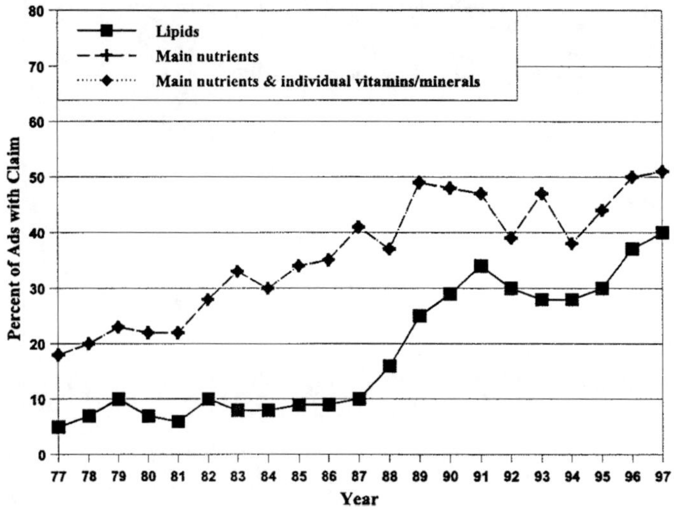

Table 7-3 Percentage of Advertisements with Claims for One or More Lipid Dimensions[2]

Number of Lipid Dimensions in Ads	1977		1983		1991		1997	
3	0.3	1.1	0.0	2.5	2.2	20.1	0.7	5.0
2	0.8		2.5		17.9		4.3	
1	3.7		5.5		14.3		34.5	
0	95.1		92.0		65.6		60.5	

Notes. [1] *Lipid dimensions* and *Main nutrients* defined in Table 7-1.
[2] Lipid dimensions are fat, saturated fat, polyunsaturated fat, monounsaturated fat, and cholesterol.

Main Nutrient Dimensions

Figures 7-2, 7-3, and 7-4 also illustrate the use of *Main nutrient claims* in advertising throughout the period with and without the breakout of individual vitamins and minerals. The evidence largely parallels that for lipids, with a few differences. Figure 7-2 illustrates that the growth in the use of main nutrient claims begins in 1982, earlier than the growth for lipids. These main nutrient claims peak in 1991, as with lipids, at 1.15 dimensions per ad, and drop in the early 1990s, stabilizing at approximately 0.9 dimensions per advertisement, a 22 percent drop from the peak. Also note that breaking out individual vitamin claims, as opposed to treating them as a summary measure, has a small effect on these data. Most multiple claims involve the main nutrients listed in Table 7-1 rather than individual vitamin claims.

As with lipids, movement in the overall mean is due more to changes in the number of nutrient dimensions featured in advertisements than to changes in the number of advertisers making any nutrient claims at all. As shown in Figure 7-3, the mean number of main nutrient dimensions in advertisements, for those advertisers making claims, rises sharply in the late 1980s and decreases substantially in the 1990s. Both changes are more pronounced than for lipids alone. By 1997, the number of nutrients in the average ad making claims has returned to the level of the mid-1980s, a drop of approximately 33 percent from the peak in 1992 in both measures.

This pattern of changes is illustrated in more detail in Table 7-4, which shows the distribution of multiple dimension advertisements for the years 1977, 1983, 1991, and 1997. In 1977 a few advertisements show copies of the nutrition label. With that exception, the inclusion of multiple dimensions in advertising does not really grow until the mid-1980s, and by 1991, 19.9 percent of all food advertising has 3 or more main nutrients mentioned explicitly in advertising, evidence consistent with the hypothesis that competitive pressure leads to greater information unfolding. After passage of the NLEA rules, the percentage of advertisements making claims for multiple dimensions falls substantially, so that by 1997 only 8.5 percent of food ads mention three or more nutrients in their advertising, less than half the 1991 rate. A much more substantial portion of the advertising in 1997 is single nutrient advertising, and as we saw in Chapter 4, much of that is for total fat.

Taken together these results are consistent with the hypothesis that the less restrictive regulatory environment of the late 1980s and the resulting increase in competitive focus on nutrition led advertisers to highlight more nutritional dimensions of their products than they had earlier. Since the NLEA, advertisers who make nutrient claims have reduced the number of dimensions included in advertising back to the levels of the mid-1980s.

Table 7-4 Percentage of Advertisements with Claims for One or More *Main Nutrients*[1]

Number of Main Nutrients in Ads	1977		1983		1991		1997	
7+	0.5	5.2	0.0	4.0	0.4	19.9	0.0	8.5
6	1.3		0.2		1.2		0.5	
5	0.2		0.4		2.2		0.2	
4	0.3		0.4		6.2		2.7	
3	2.9		3.0		9.9		5.1	
2	2.6		7.3		12.5		14.2	
1	9.9		21.9		14.5		28.4	
0	82.3		66.8		53.1		48.9	

Notes. [1] *Main Nutrients* listed in Table 7-1.

Is There Competition Among "Bads?" The Case of Fats and Oils

The unfolding hypothesis implies that firms with a *relative* advantage over their competitors will be led to advertise that advantage whenever profitable. An implication of this theory is that even advertisers in "bad food" categories constrained to make only truthful claims may be induced to make nutrition and other health claims, to highlight differences *within* the food category that make some choices nutritionally superior to others. Thus, if producers get a new opportunity to highlight their advantages, they should react to the change with new advertising claims.

The advertising data here provides an opportunity to test the unfolding theory as it relates to competition on bads. The *Fats and Oils Category* is defined in our data to include any butter, margarine, spread, lard, shortening, oil, or related product. These products are all high fat products, or substitutes for such products, and thus, the category is generally considered to be a "bad food" category. However, fat and oil products vary considerably in the *type* of fat they contain and in the amount of fat per serving. Thus, within the category there is considerable variation in the health implications of using the different products.

For most of the period examined here, saturated fats are considered to be the type of fat most hazardous to health. Saturated fats increase the risk of cardiovascular disease, while polyunsaturated and monounsaturated fats generally do not.[9] Substituting unsaturated fats for saturated fats is expected to reduce the risk of cardiovascular disease, and thus under the unfolding theory, producers of lower saturated fat products within the category have a health advantage that they would be expected to promote if the regulatory rules allow such competition.

To assess the extent of competition of this type, we must also control for key regulatory constraints. Recall that key regulatory events related to health claims are listed in Table 6-1. We again focus on the four major events used in the regression analyses in Chapter 6. Of particular note for the fats and oils category, the final FDA labeling rules prohibit heart claims for all products that are not "low" in fat, as defined in the regulations. Thus, the NLEA-based rules prohibit explicit health competition on the *type* of fat in fat and oil products, though lipid claims without any health context would still be allowed. Moreover, low fat substitutes for higher fat products that do not have nutrition value, as defined by the regulations, are also barred from making health claims.[10]

Thus, focusing on these regulatory events, if the unfolding theory applies even in bad food categories, health claims for fats and oils would be expected to be minimal prior to 1983 (or possibly 1987, if the labeling rules are implicitly binding on advertising), to increase

---

[9] In the late 1980s and early 1990s, scientific evidence accumulated indicating that transfatty acid, a type of unsaturated fatty acid, also increases the risk of cardiovascular disease by increasing serum cholesterol. Transfatty acid claims are not allowed on food labels under the NLEA rules during the period studied here. Recent labeling proposals could change this. However, general publicity about the scientific findings could have led to reduced margarine consumption in the 1990s, because transfatty acids are most common in those products within the category. The new evidence could also affect marketing of the products within the category. However, if producers assumed that the labeling restrictions that limit the ability to discuss the relative benefits of different fats applied to advertising as well, these responses would be limited.

[10] This provision, known as the *jelly bean rule*, is designed to prevent highly sugared but low fat products from making health claims. Perhaps unintentionally, the provision also prohibits health claims for other nonfat products, such as nonfat substitutes for fats in cooking or nonfat salad dressings.

substantially at this point and then to fall rapidly following the FDA proposals in 1990, and probably to be eliminated under the final label rules in 1993 and the coordinating FTC policy in May 1994. Nutrient content claims dealing with fat composition would presumably parallel these health claims, but they might persist in the post-NLEA period, even if health claims are prohibited, if consumer knowledge of the health issue is sufficiently strong by that point.

Figure 7-5 illustrates the percent of fat and oil advertisements in each year making an explicit disease claim (heart disease in this case) as well as the percent making disease or affiliated claims (mostly serum cholesterol claims, together with the heart claims).

Focusing first on explicit disease claims, the evidence indicates that virtually no disease claims are made through 1983, the year the FTC formally ends its Food Rule proceeding and publicly states that it will judge disease claims under a deception standard. Disease claims begin rising in 1984, rise more steeply after 1987, and by 1990, 33.3 percent of all fat and oil ads include specific disease claims. These are primarily lower saturated fat products touting their heart healthiness. Immediately following the FDA actions in 1990 and 1991, the use of disease claims falls to 2.7 percent of ads in 1993 before falling to zero in 1994, the year of the FTC statement. This evidence is quite consistent with the unfolding theory as described above: the use of disease claims increases rapidly following the lifting of the regulatory restrictions and falls dramatically when the prohibition is reinstituted under the NLEA rules. During the period when disease claims are allowed, the reasons to choose one fat over another become a dominant focus of competition in the fats and oils category.

Figure 7-5 Percentage of Fats and Oils Advertisements with *Disease or Affiliated Claims*[1]

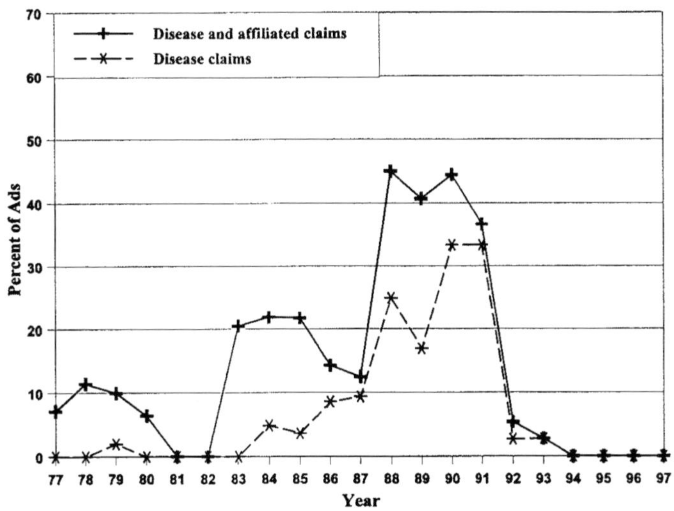

Figure 7-6 Projections from Probit-2 Model for Probability of *Disease or Affiliated Claim*[2] in Fats and Oils Advertising, Monthly Data

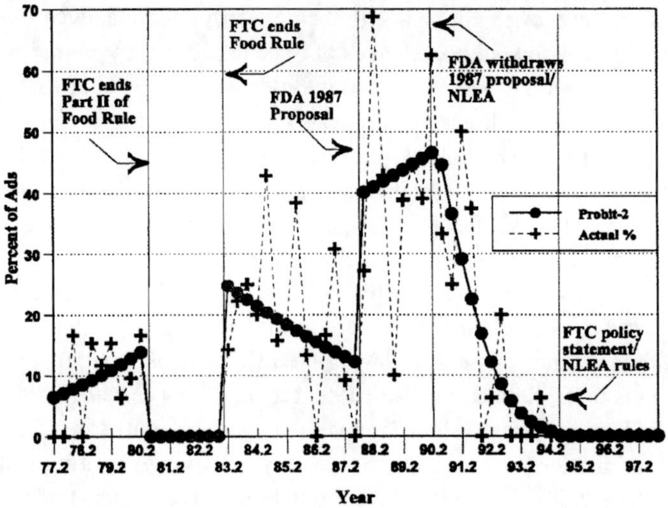

Notes. [1] The Probit-2 estimate is given in Table 7-5 and relates key regulatory events to use of disease and affiliated claims.

[2] In fats and oils advertising these claims are virtually all serum cholesterol or heart claims.

Figure 7-5 also gives the percentage of ads that have either an affiliated or a disease claim. The evidence is quite similar. Since affiliated claims are not so clearly prohibited even in the 1970s, they carry a somewhat lower regulatory risk for advertisers. This lower risk is evident in the greater use of affiliated claims in the late 1970s, when between 6.5 and 11.4 percent of fat and oil advertising include serum cholesterol claims. These claims fade as the FTC Food Rulemaking reaches a decision point on heart-related claims and rise dramatically and immediately to more than 20 percent of ads in 1983, once the FTC votes to end its Food Rulemaking and to allow health claims under normal deception rules. By 1988, 45 percent of fat and oil ads include serum cholesterol or heart claims. These claims remain an important feature of marketing in the category until 1992, when they fall from 36.7 percent in 1991 to 5.4 percent of advertising in 1992 following the publication of the proposed NLEA labeling rules that prohibited health claims for fat and oil products and to 0.0 in 1994 when those rules are effective. Health claims do not reappear in fat and oil advertising through the end of our sample in 1997.

To provide an assessment of the statistical significance of these changes, we estimate simple time series regressions using the individual data for each fat and oil advertisement and a series of dummy variables for the key regulatory events. The dependent variable is a dummy variable indicating whether the ad has a disease or affiliated claim. The regression specifications are comparable to those used in Chapter 6 for the broader analysis. Our sample includes 720 fat and oil advertisements.

Table 7-5 *Disease and Affiliated Claim* Regression Results for Fats & Oils

Variable	Linear-1	Linear-2	Probit-1	Probit-2
Constant	0.104**	-1.80	-1.26**	-12.60
	(3.31)	(-0.66)	(8.33)	(0.90)
$D_{5/1980}$	-0.104**	-0.14	—[1]	—[1]
(FTC ends Part II of Food Rule)	(-2.11)	(-1.37)		
$D_{1/1983}$	0.183**	0.25**	0.36*	0.02
(FTC ends entire Food Rule)	(4.02)	(2.87)	(1.93)	(0.02)
$D_{8/1987}$	0.253**	0.28**	0.74**	0.93**
(FDA health claim proposal)	(6.28)	(3.29)	(4.83)	(2.68)
$D_{2/1990}$	-0.278**	-0.05	-0.84**	0.13
(FDA withdraws 1987 proposal/NLEA)	(-6.28)	(-0.58)	(4.80)	(0.40)
$D_{5/1994}$	-0.159**	0.11	—[1]	—[1]
(FTC policy statement; FDA/NLEA rules effective)	(-2.98)	(1.10)		
Time	—	0.02	—	0.14
		(0.70)		(0.81)
Time*$D_{5/1980}$	—	-0.02	—	—[1]
		(-0.40)		
Time*$D_{1/1983}$	—	-0.03	—	-0.25
		(-0.54)		(-1.29)
Time*$D_{8/1987}$	—	0.06	—	0.18
		(1.22)		(1.05)
Time*$D_{2/1990}$	—	-0.15**	—	-0.69**
		(-3.06)		(-3.20)
Time*$D_{5/1994}$	—	0.12**	—	—[1]
		(2.51)		
Adj. R-squared	0.131	0.151		
Log-Likelihood			-274.14	-262.85
n	720	720	568	568

Notes. t-statistics in parentheses. Dependent variable is a dummy variable indicating that an ad has claim.

[1] The probit model could not be estimated for June 1980-January 1983 and for June 1994-October 1997 because no ads during these periods made disease or affiliated claims, thus producing no variation in the data during these periods. The sample size, n, reflects the more limited time span used in the estimates.

The column labeled *Linear-1* in Table 7-5 presents the results for the simplest linear model with discrete dummy variables for the regulatory events. These results indicate that the shifts between regimes are statistically significant in all cases and that an average of 10.4 percent of ads have disease and affiliated claims prior to the FTC decision in May 1980. All fats and oils advertisers stop making heart-related claims after this decision until the FTC votes in late December 1982 to adopt a case-by-case deception approach to the claims. Disease and affiliated claims immediately rise, averaging 18.3 percent of all fats and oils advertising in the months between that decision and August 1987, when the FDA proposes adopting a similar approach for labeling. The model indicates that an additional 25.3 percent of fats and oils advertising also includes these claims during the period following this decision, for a total of 43.6 percent on average. Following the FDA retraction of this proposal in favor of a more regulatory approach in February 1990, advertising using disease and affiliated claims falls by 27.8 percentage points in the category. Finally in the period following May 1994, the remaining 15.9 percent of advertising also stops using the claims.

These estimates suggest that the regulatory rules had statistically significant and sizable effects on the use of disease and affiliated claims in the fats and oils category. The estimates also suggest that producers actively focused on the heart health implications of fat composition in their advertising when free to do so under the regulatory rules.

The column labeled *Linear-2* elaborates on the basic model by allowing an underlying trend (to reflect trends in scientific knowledge and general information diffusion) and trend interactions with the regulatory dummies. The model provides a better fit for the data but has qualitatively similar results. The estimate indicates no significant time trend[11] and significant shifts in the last four regulatory events.

Comparable probit models are also estimated, but because no advertisements have any disease or affiliated claims between February 1980 and February 1983 and after February 1994, the model is estimated only for the remaining periods.[12] The projections from the models are very similar to those of the comparable linear model. For instance, Figure 7-6 illustrates the projections for the months in our sample from the *Probit-2* model, together with the actual percent of advertisements with a disease or affiliated claim in each month of the sample. The regulatory events are also indicated on the graph. The projections from the *Linear-2* model would be very similar. The graph illustrates that despite considerable month to month variation in the use of disease and affiliated claims, a clear and substantial difference in the pattern of use is apparent across regulatory periods paralleling the pattern seen in the simple *Linear-1* model described above. Moreover, the figure illustrates that the reactions to substantial changes in the regulatory rules are sizable and occur relatively quickly both in reducing claims when the rules are tightened and in increasing claims when restrictions are lifted. This evidence is consistent with the view that advertisers react strongly to the regulatory rules by making specific claims in their ads when issues are important within the category, and thus, that the information content of the advertising is important to them.

Nutrient claims are also used for this class of products. Figure 7-7 illustrates the percent of fat and oil advertising that makes saturated fat claims during this period – the primary basis for heart-related claims. The percent of ads that includes a saturated fat claim rises steadily through 1990, when it peaks at 52.8 percent of advertising in the category. The use of saturated fat claims begins to fall at this point, dropping to 42.9 percent of advertising by 1996 and 8.3 percent in 1997, but this drop clearly lags behind the reduction in the use of health claims. Thus, the data suggests that consumer interest in the health issues implicit in choosing fats and oils does not fall dramatically in 1990, and thus, that the regulatory constraints may indeed be the cause of the precipitous decline in the use of health claims in

[11] The scientific basis for the relationship between saturated fat and heart disease is relatively well established even at the start of our period. See Ippolito and Mathios (1996), Pappalardo and Ringold (2000), or National Research Council (1989).

[12] Specifically, the data during these periods has to be dropped and the model estimated relative to the initial period. When the predictions of the probit model are projected over time, they coincide very well with the predictions of the corresponding linear models.

the category. Moreover, in periods when health claims are restricted, firms generally use nutrient claims (saturated fat claims in this case) to reach out to consumers who understand the importance of the nutrient.

Finally, Figure 7-8 illustrates the actual number of advertisements per year in our sample for fats and oils. The level of advertising for the category shows a small positive trend through 1990, but this trend is reversed after that point so that by 1997, fat and oil product advertising has fallen to 43 percent of its 1977 level and to 20 percent of its peak in 1989. The figure also illustrates the number of fat and oil ads in our sample that include a disease or affiliated claim and those that make a saturated fat claim. The data clearly illustrate that saturated fat claims and disease and affiliated claims are highly correlated prior to 1990, as expected. The data also indicate that after health claims are prohibited in this category, claims about the saturated fat content of fats and oils begin to fall as does all advertising in the category. By 1997, few fat and oil producers seem to be competing on the nutritional characteristics of their products.

Figure 7-7 Percentage of Fats and Oil Advertisements with *Saturated Fat Claims*

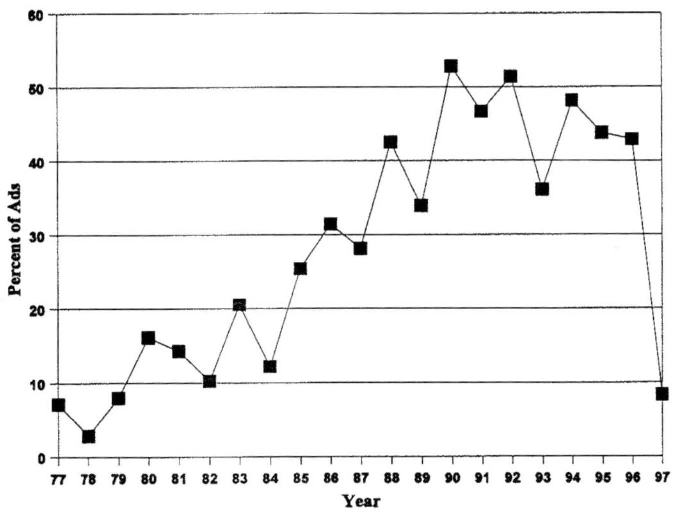

Figure 7-8 Number of Fats and Oils Advertisements per Year and Number with *Saturated Fat, Cholesterol,* and *Disease or Affiliated Claims*

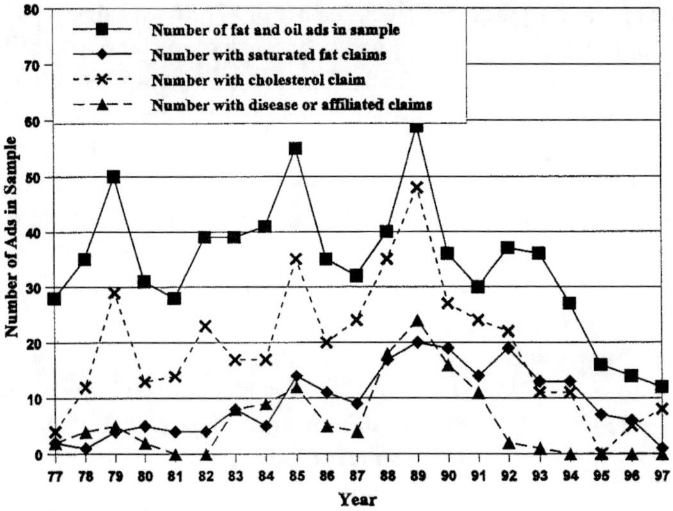

Taken together this evidence indicates that competition on *bads* does occur and can become a major focus of competition in a particular category, as in the fats and oils category here. Having less of a bad is of course a good thing, and apparently advertisers believe that they can communicate these differences to consumers in advertising in a way that enhances their products' sales.

Moreover, the evidence in our sample indicates that the fat and oil category is the food category where disease and affiliated claims appear first as a significant feature of advertising (see Table 5-1, for instance). Why these health claims would be used first in such a high fat category is an open question. Possibly the general discussion of fats and heart disease from public health and general information sources sufficiently sensitized consumers to fats as a health concern that producers view it as an easy and salient message to communicate. Alternatively, because the amount of fat in these products is relatively large and fat is used as an input to many foods prepared at home, shifting across types of fat produces sizable health benefits, and thus, represents a significant marketing opportunity. Whatever the rationale, the evidence indicates that the use of health claims focusing on heart-related differences in the amount and type of fat is an early and significant feature of competition in the fats and oils category prior to the adoption of the NLEA rules.

ADVERTISING AND BROADER AUDIENCES: DO PRODUCERS REACH OUT WITH NEW INFORMATION?

In exploring the role of advertising in markets, one of advertising's possible strengths is its potential to reach out to consumers with information. When there is information to convey that would affect consumer behavior in ways that would be profitable to advertisers,

advertisers have strong incentives to bring that information to the audiences who would use it. In earlier work on the cereal market (Ippolito and Mathios, 1989) and on fat consumption (Ippolito and Mathios, 1996) this incentive to reach broader audiences was discussed as a key reason for the expectation that diets would change as regulatory rules allowed firms to make explicit disease claims in advertising. Unfortunately the production and consumption data available in those studies do not allow a direct test of this economic theory, but only a test of its implications on behavior; diets did improve when the restrictions on health claims were lifted.

The advertising data available in this study allows us to examine this theory more directly. Magazine food advertising is primarily placed in "women's magazines," reflecting the primary food shopper in many households. This dictates the choice of our sample to include the 5 leading women's magazines, as described in Chapter 2, namely *Better Homes & Gardens*, *Good Housekeeping*, *Ladies' Home Journal*, *McCalls*, and *Women's Day*. Three high circulation "general readership" magazines that carry some food advertising are also included in the sample, *Time*, *Newsweek*, and *Reader's Digest*. In 1977 the 5 women's magazines in our sample contain 558 food advertisements compared to 59 food ads in the 3 general readership magazines, illustrating the relative dominance of the women's magazines as food advertising vehicles.

If the ability to make disease and affiliated claims allows firms to bring information to consumers in new ways, we might expect firms to bring that information to their normal target audience, but also to reach out to other possible consumers as well, as they use the new information to attempt to expand the demand for their products. Figure 7-9 illustrates the percent of ads that contain a disease or affiliated claim in each class of magazines. As the regulatory constraints are lifted in the mid-1980s and again after the NLEA rules are put in place in the 1990s, the use of disease and affiliated claims rises in the women's magazines, but rises considerably more in the general readership magazines. In 1989 at the peak, 20.9 percent of all food ads in our general readership magazine sample contained a disease or affiliated claim compared to 6.6 percent of ads in the women's magazine sample. In 1983 and in 1993, before and after the peak, both samples have disease and affiliated claims in less than two percent of advertisements. Thus, both samples begin and return to the same level of claim use when the regulatory environment is not receptive to these claims, but the rise is substantially greater in the general readership magazines when the regulatory environment does not prevent these claims. The post-NLEA rise after 1995 follows the same pattern.

The theory that producers reach out to the broader audience with health information is also supported by data on the number of ads in the two types of magazines. As shown in Figure 7-10, the number of food ads in women's magazines has been trending down since the mid-1980s. In contrast, at the height of the health claim period in 1989, the number of food ads in the general readership magazines increases substantially to 140 percent of its 1977 level before returning to its 1977 level in 1993 when the use of health claims falls. The post-NLEA data are also consistent with this hypothesis, except for 1997, when general readership ads fall.

Figure 7-9 Percentage of Food Ads with *Disease and Affiliated Claims* by Magazine Type[1]

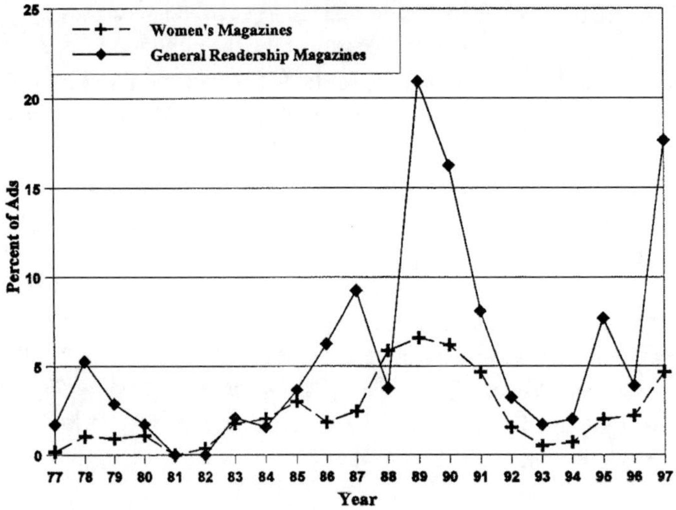

Figure 7-10 Number of Food Ads by Magazine Type as a Percent of 1977 Level[2]

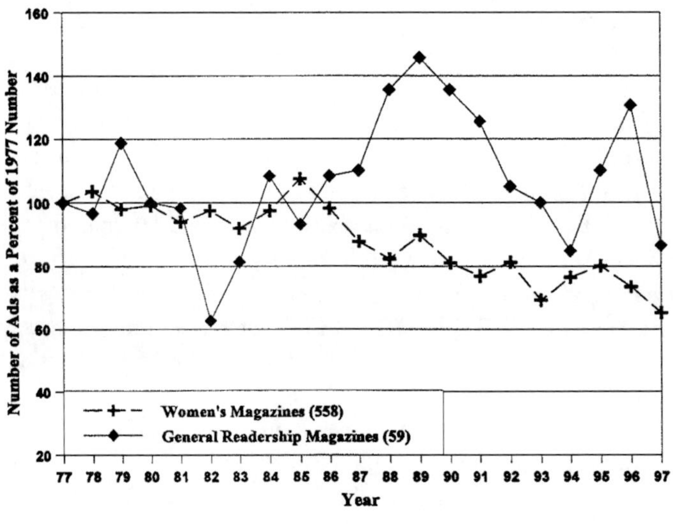

Notes. [1] General readership magazines in the sample are Time, Newsweek, and Reader's Digest. Women's magazines are Better Homes & Gardens, Good Housekeeping, Ladies' Home Journal, McCalls, and Women's Day.

[2] The number of ads in each sample in 1977 is given in parenthesis in the legend.

Finally, Figures 7-11 and 7-12 illustrate the same phenomenon for fat and saturated fat claims. These claims often accompanied heart or other health claims, but they are also used in isolation. The data indicate that reaching out disproportionately to the general readership audience occurs only during or immediately after the intensive health claims periods when producers had information about health effects that could be conveyed to potential consumers.

Taken together these data are generally consistent with the hypothesis that advertisers will attempt to spread information that expands the demand for their products to broader audiences when allowed to do so.

SUMMARY AND CONCLUSION

This chapter examines several predictions of economic theories of advertising's role in markets. First and foremost, the evidence here suggests that most magazine food advertising contains a variety of specific informative claims. In this forum, advertising provides information. As regulatory rules have been relaxed and general awareness of nutritional issues has increased, specific nutrition claims have become a major focus of food advertising. In the 1990s approximately half of all food ads have specific nutrient claims. Claims about other search and experience characteristics of foods, such as available flavors or varieties, suggestions for using the product, convenience features, new or changing products, and taste or other sensory characteristics, are all common features of food advertising. Taken together, the evidence suggests that specific claims about search and experience characteristics of foods are a dominant feature of food advertising in our sample.

Figure 7-11 Percentage of Ads with *Total Fat Claims* by Magazine Type[1]

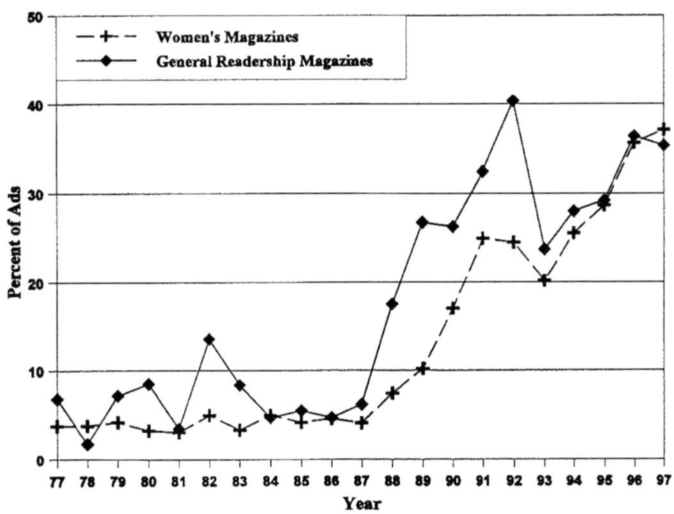

Figure 7-12 Percentage of Ads with *Saturated Fat Claims* by Magazine Type[1]

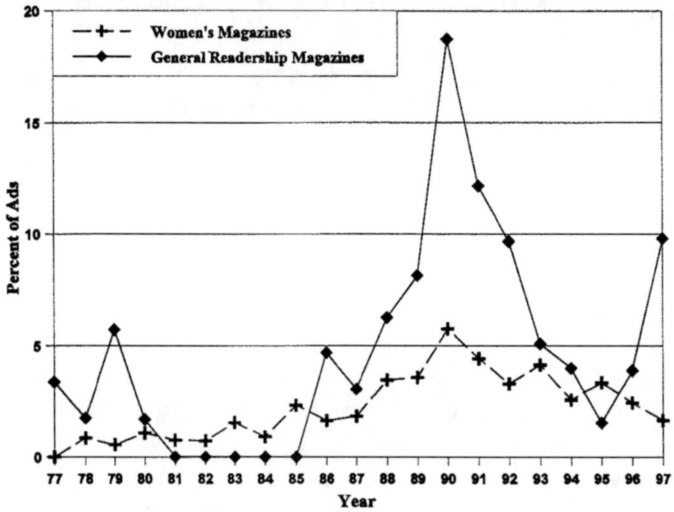

Notes. [1] General readership magazines in the sample are *Time, Newsweek,* and *Reader's Digest.* Women's magazines are *Better Homes & Gardens, Good Housekeeping, Ladies' Home Journal, McCalls,* and *Women's Day.*

Perhaps the high level of informative-type claims is a reflection of the medium we examine – magazine advertising. Becker and Murphy (1993) predict that print media will have a larger fraction of informative advertising than TV and radio, because the latter media must compensate consumers (with free programming) to accept the type of advertising presented there. We have no advertising from these TV and radio media with which to assess the relative levels of specific informative claims, but clearly the level of informative claims could vary with the medium.

The evidence also suggests that competitive pressure generated by the use of health claims leads to considerable information unfolding relative to the periods when health claims are prohibited or after the NLEA rules are in place. The average number of nutrients mentioned in ads doubles from the early 1980s to 1990, but then drops by more than 20 percent by 1997. More telling, perhaps, is the evidence on the proportion of ads with multiple claims, which follows an even steeper rise as health claims increase but then declines significantly in the post-NLEA period. In 1991, at the height of the health claim advertising, 20 percent of food advertisements have claims for 3 or more different nutrients, in 1983 only 4 percent of ads highlight 3 or more nutrients and in 1997 only 8.5 percent do. Thus, the evidence is consistent with the hypothesis that a more open environment leads to competitive pressures that induce producers to reveal information on more nutrient dimensions in advertising.

The evidence from the fats and oils market clearly illustrates that advertisers will compete on the relative health features of foods even in "bad" food categories if differences are sufficiently large. Fats and oils producers are among the earliest and heaviest users of health claims as they advertised how the choice among fat products is related to heart disease risk. This advertising has disappeared under the NLEA rules that do not allow foods not considered "good" foods to mention the health issues implicit in the choice of these products. Thus, the NLEA rules may have shifted the focus of competition in the category away from nutrition to other issues.

Finally, the evidence suggests that producers do reach out to potential customers when they have information that would increase demand for their products, as predicted by theory. When health claims are allowed, advertisers significantly increase the number and health focus of food advertisements in general readership magazines, thus apparently reaching out beyond their normal magazine media choices to the mass audience expected to find this information useful.

Chapter 8

CONCLUSION

Advertising can be an important source of information for consumers. Advertisers have incentives to highlight product features valued by potential customers. As long as market mechanisms, or enforcement of deceptive advertising laws, sufficiently discourage deceptive claims, the pressures created by advertising competition should push producers to improve their products in dimensions that consumers value and should improve the information environment in which consumers make product choices

This report examines the types of claims made in magazine food advertising during the years 1977 to 1997. The report uses original data collected from a large-scale, systematic sample of the leading women's and general readership magazines. Besides providing a wealth of information on the content of food advertising, we also attempt to better understand two issues: how economic forces affect food advertising over time, especially as they relate to nutrition and health claims, and how producers' incentives to focus on nutrition and health vary under the different policies adopted during these years, including those adopted after the *Nutrition Labeling and Education Act of 1990.*

The data show that nutrition-related claims have become a major feature of food advertising and an important focus of competition. In the broadest sense, the evidence here shows some reduction in most classes of nonnutritive claims and an increase in specific nutrition-related claims during these years. For instance, taste, aroma, and/or texture claims are made in more than 80 percent of ads through the late 1970s and early 1980s, but fall to 65 percent of ads by 1997. In contrast, 28 percent of ads in 1977 contain a specific nutrition-related claim of some type; by 1990, 55 percent of ads have such claims, a level that is approximately maintained during the 1990s. Thus our data indicate that the types of information in advertising can shift markedly, depending on what producers choose to feature in their advertising.

The data also indicate that regulatory rules and enforcement policy matter; firms move away from nutrition or health claims when regulatory risks rise and increase the use of these claims when regulatory risk falls. As the FTC pursued its proposed Food Rule in the late 1970s and early 1980s, producers reduced their use of claims challenged in the rulemaking. After the FTC made decisions to end various parts of the rulemaking in favor of a case-by-case policy of pursuing deceptive claims, advertisers again began to highlight nutrition and diet-disease issues in advertising.

A similar pattern is found as various decisions are made leading to the current post-NLEA environment. For instance, disease and affiliated health claims increase significantly following the FDA's proposed adoption of a "reasonable basis" standard in the late 1980s, to a peak of 8.9 percent of ads by 1989. They fall substantially after the agency reverses that position in 1990, to under 1 percent of ads in 1993. By 1997, after the final NLEA rules are in place, disease and affiliated health claims returned to 72 percent of their 1989 peak level. The use of health claims in various product categories varies with the requirements of the NLEA labeling rules, presumably reflecting perceived enforcement risk in advertising.

Focusing more directly on the changes in the post-NLEA period, several findings are worth noting. The nutritional focus in advertising has narrowed substantially in the post-NLEA period. Total fat has become the primary nutritional basis of advertising competition, away from other major nutrients, such as saturated fat, cholesterol, and sodium that had been important before the NLEA. The number of advertisements that make comparative claims about nutrients has also dropped sharply, to very low levels for nutrients other than total fat.

The most dramatic change in the use of health claims in the post-NLEA period occurs for fat and oil products, such as cooking oils, margarines, and related products. Prior to the NLEA, we observe vigorous competition on the fat composition of various products. In the late 1980s, nearly half of all advertisements for fat and oil products focus on the heart implications of fat choices. In contrast, after the 1993 NLEA rules are in place, no ads mention the health reasons to choose unsaturated fats over saturated fats. The amount of advertising also falls dramatically in the post-NLEA period, as fat and oil producers no longer compete aggressively on the nutritional and health implications of choosing one product over another.

Advertising for "good foods" also does not increase in the NLEA period. Advertising falls significantly for fruits and vegetables after 1990, and health claims are made for fewer products in the category. Advertising also falls or remains stable in other "good food" categories.

Certainly some of these advertising changes were explicitly envisioned in the NLEA labeling rules. The elimination of a health focus in the fats and oils category was designed into the rules. However, other changes may not have been anticipated, as with the reduced use of comparative claims. Whether these changes serve consumer interests is an important topic for further attention.

Regarding the economic theories of advertising, the evidence here provides considerable support for the view that advertising is an important source of information. The evidence documents that most food advertisements in magazines include specific claims about both nutritional and nonnutritional features of the products. In fact, most ads include a number of product characteristic claims. During the period when the regulatory environment is most open in the late 1980s, the nutritional competition among producers appears most intense, leading to greater unfolding of nutrient claims in advertising.

The ultimate question of which regulatory and legal policies best serve consumer interests requires that we relate the advertising changes described here to consumers' food choices. Until that work is done, this study provides us with a much needed part of that evaluation: objective and detailed information on the actual content of food advertising during the different policy periods that characterize 1977 to 1997.

Marketing is often controversial. Producers are trying to sell their products. But marketing claims about important product characteristics – subject to market and enforcement limits on deception – unleash competitive forces that play an important role in shaping the mix of products available in the market and in attracting consumers to products with desired characteristics. As science has shown the importance of nutrition in disease risks, advertising has focused increasingly on nutritional characteristics of food. In crafting policy that serves consumers' interests, it is important that we understand the role of marketing in consumer goods settings. We hope this evidence contributes to that effort.

APPENDIX A.
COMPUTERIZED CODING INSTRUMENT

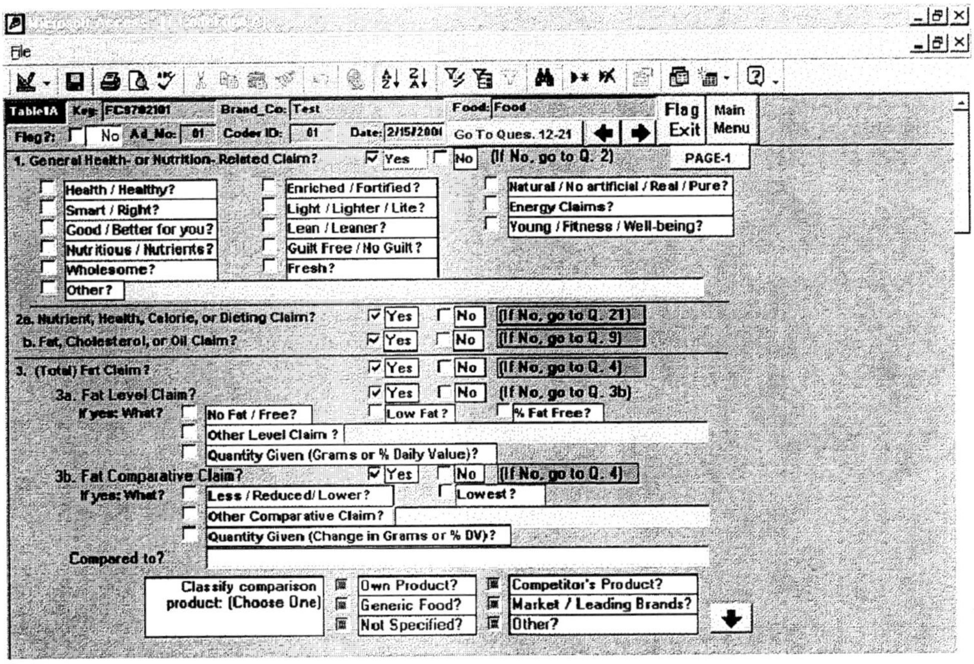

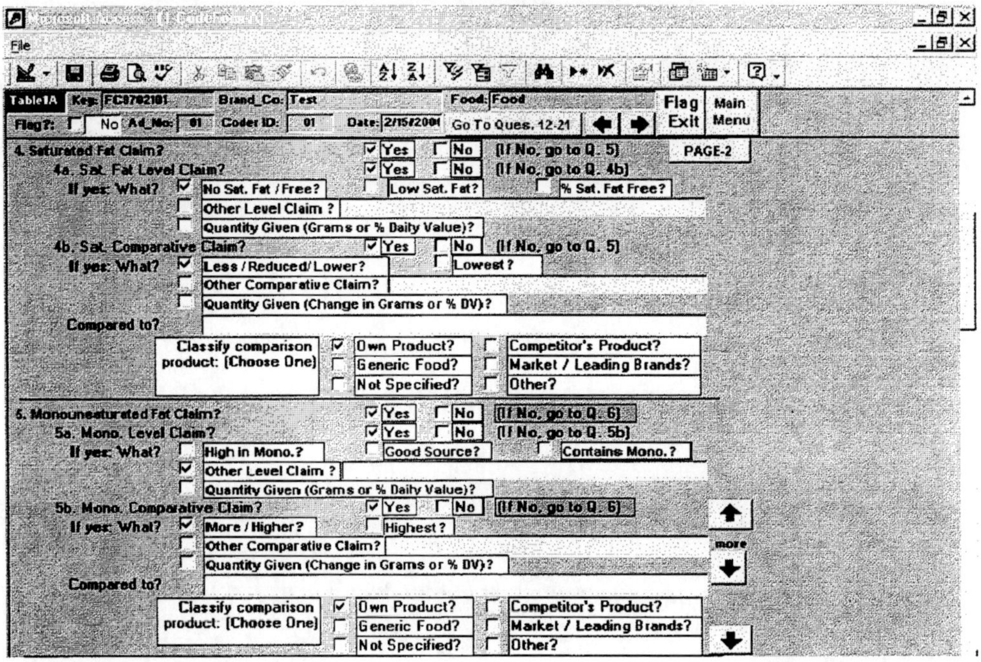
File
Table1A
Key: FC9702101
Brand_Co: Test
Food: Food
Flag Exit
Main Menu
Flag?: No
Ad_No: 01
Coder ID: 01
Date: 2/15/2000
Go To Ques. 12-21
4. Saturated Fat Claim? Yes No (If No, go to Q. 5)
PAGE-2
4a. Sat. Fat Level Claim? Yes No (If No, go to Q. 4b)
If yes: What? No Sat. Fat / Free? Low Sat. Fat? % Sat. Fat Free?
Other Level Claim ?
Quantity Given (Grams or % Daily Value)?
4b. Sat. Comparative Claim? Yes No (If No, go to Q. 5)
If yes: What? Less / Reduced/ Lower? Lowest?
Other Comparative Claim?
Quantity Given (Change in Grams or % DV)?
Compared to?
Classify comparison product: (Choose One)
Own Product? Competitor's Product?
Generic Food? Market / Leading Brands?
Not Specified? Other?
5. Monounsaturated Fat Claim? Yes No (If No, go to Q. 6)
5a. Mono. Level Claim? Yes No (If No, go to Q. 5b)
If yes: What? High in Mono.? Good Source? Contains Mono.?
Other Level Claim ?
Quantity Given (Grams or % Daily Value)?
5b. Mono. Comparative Claim? Yes No (If No, go to Q. 6)
If yes: What? More / Higher? Highest?
Other Comparative Claim?
Quantity Given (Change in Grams or % DV)?
more
Compared to?
Classify comparison product: (Choose One)
Own Product? Competitor's Product?
Generic Food? Market / Leading Brands?
Not Specified? Other?

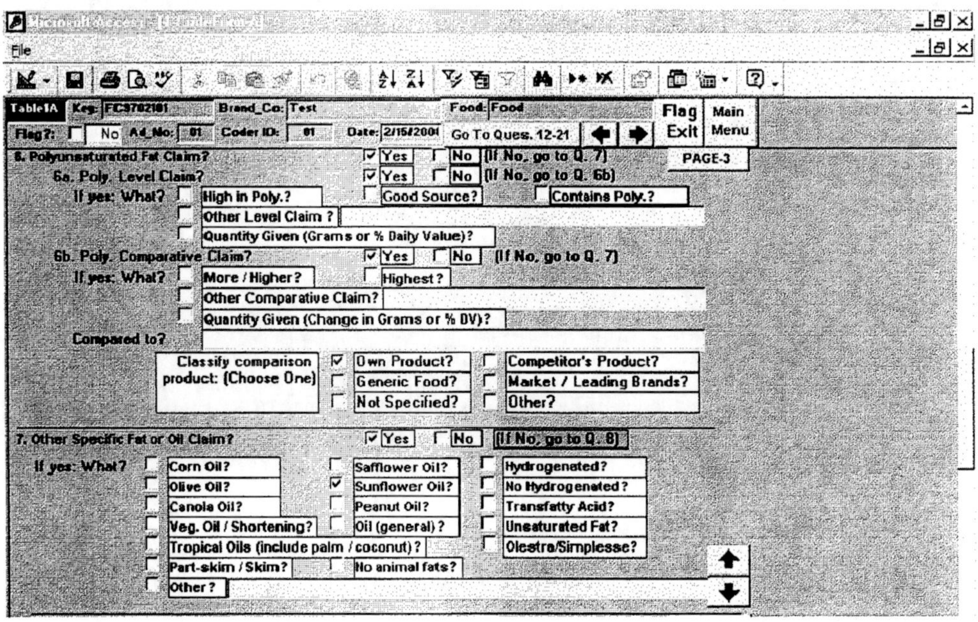
File
Table1A
Key: FC9702101
Brand_Co: Test
Food: Food
Flag Exit
Main Menu
Flag?: No
Ad_No: 01
Coder ID: 01
Date: 2/15/2000
Go To Ques. 12-21
6. Polyunsaturated Fat Claim? Yes No (If No, go to Q. 7)
PAGE-3
6a. Poly. Level Claim? Yes No (If No, go to Q. 6b)
If yes: What? High in Poly.? Good Source? Contains Poly.?
Other Level Claim ?
Quantity Given (Grams or % Daily Value)?
6b. Poly. Comparative Claim? Yes No (If No, go to Q. 7)
If yes: What? More / Higher? Highest?
Other Comparative Claim?
Quantity Given (Change in Grams or % DV)?
Compared to?
Classify comparison product: (Choose One)
Own Product? Competitor's Product?
Generic Food? Market / Leading Brands?
Not Specified? Other?
7. Other Specific Fat or Oil Claim? Yes No (If No, go to Q. 8)
If yes: What? Corn Oil? Safflower Oil? Hydrogenated?
Olive Oil? Sunflower Oil? No Hydrogenated?
Canola Oil? Peanut Oil? Transfatty Acid?
Veg. Oil / Shortening? Oil (general)? Unsaturated Fat?
Tropical Oils (include palm / coconut)? Olestra/Simplesse?
Part-skim / Skim? No animal fats?
Other?

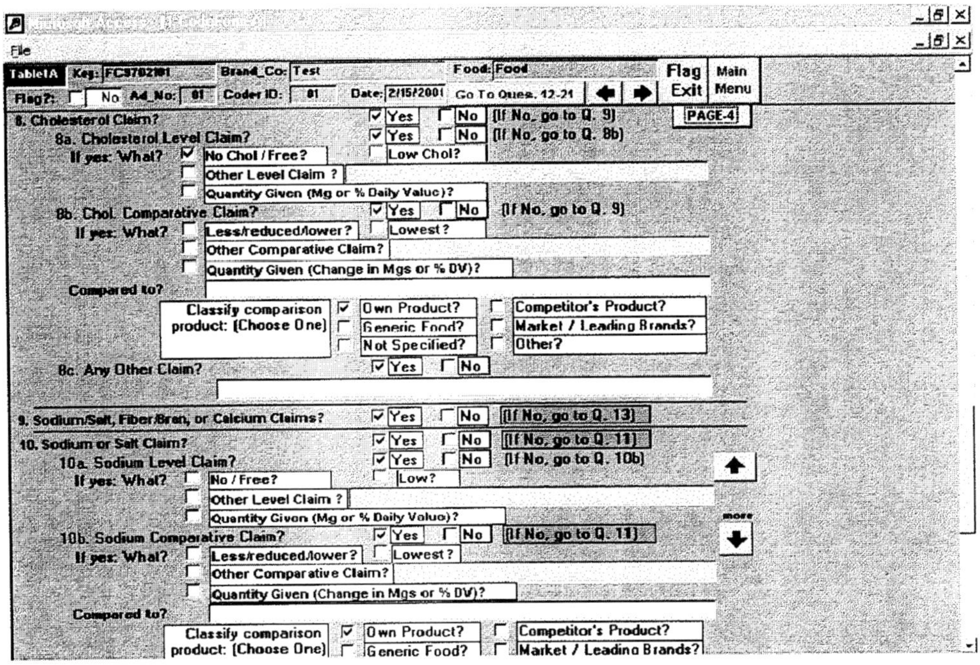
File
TableIA Key: FC9702101 Brand_Co: Test Food: Food
Flag?: No Ad_No: 01 Coder ID: 01 Date: 2/15/2001 Go To Ques. 12-21
Flag Exit
Main Menu
PAGE-4
8. Cholesterol Claim? Yes No (If No, go to Q. 9)
8a. Cholesterol Level Claim? Yes No (If No, go to Q. 8b)
If yes: What? No Chol / Free? Low Chol?
Other Level Claim ?
Quantity Given (Mg or % Daily Value)?
8b. Chol. Comparative Claim? Yes No (If No, go to Q. 9)
If yes: What? Less/reduced/lower? Lowest ?
Other Comparative Claim?
Quantity Given (Change in Mgs or % DV)?
Compared to?
Classify comparison product: (Choose One)
Own Product? Competitor's Product?
Generic Food? Market / Leading Brands?
Not Specified? Other?
8c. Any Other Claim? Yes No
9. Sodium/Salt, Fiber/Bran, or Calcium Claims? Yes No (If No, go to Q. 13)
10. Sodium or Salt Claim? Yes No (If No, go to Q. 11)
10a. Sodium Level Claim? Yes No (If No, go to Q. 10b)
If yes: What? No / Free? Low?
Other Level Claim ?
Quantity Given (Mg or % Daily Value)?
more
10b. Sodium Comparative Claim? Yes No (If No, go to Q. 11)
If yes: What? Less/reduced/lower? Lowest ?
Other Comparative Claim?
Quantity Given (Change in Mgs or % DV)?
Compared to?
Classify comparison product: (Choose One)
Own Product? Competitor's Product?
Generic Food? Market / Leading Brands?

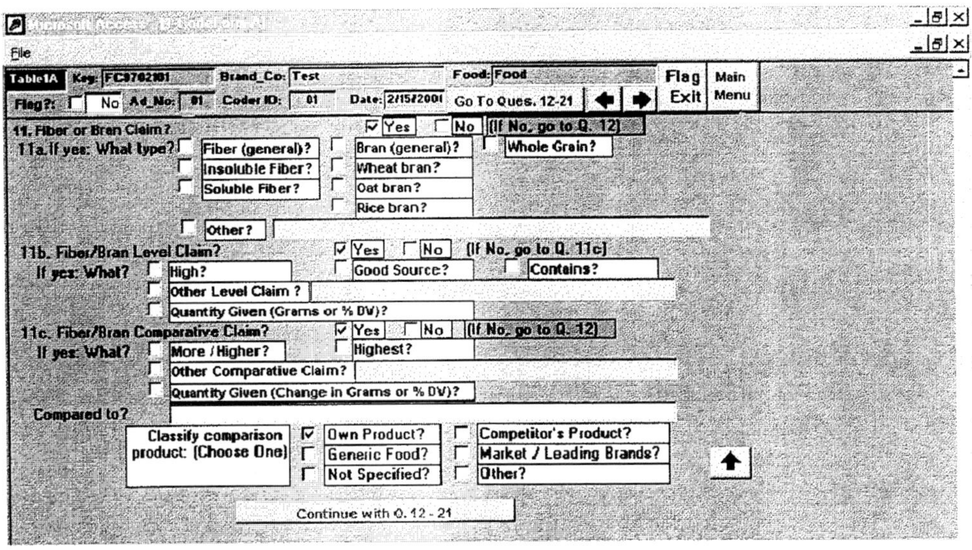
File
TableIA Key: FC9702101 Brand_Co: Test Food: Food
Flag?: No Ad_No: 01 Coder ID: 01 Date: 2/15/2001 Go To Ques. 12-21
Flag Exit
Main Menu
11. Fiber or Bran Claim? Yes No (If No, go to Q. 12)
11a. If yes: What type?
Fiber (general)? Bran (general)? Whole Grain?
Insoluble Fiber? Wheat bran?
Soluble Fiber? Oat bran?
Rice bran?
Other?
11b. Fiber/Bran Level Claim? Yes No (If No, go to Q. 11c)
If yes: What? High? Good Source? Contains?
Other Level Claim ?
Quantity Given (Grams or % DV)?
11c. Fiber/Bran Comparative Claim? Yes No (If No, go to Q. 12)
If yes: What? More / Higher? Highest?
Other Comparative Claim?
Quantity Given (Change in Grams or % DV)?
Compared to?
Classify comparison product: (Choose One)
Own Product? Competitor's Product?
Generic Food? Market / Leading Brands?
Not Specified? Other?
Continue with Q. 12 - 21

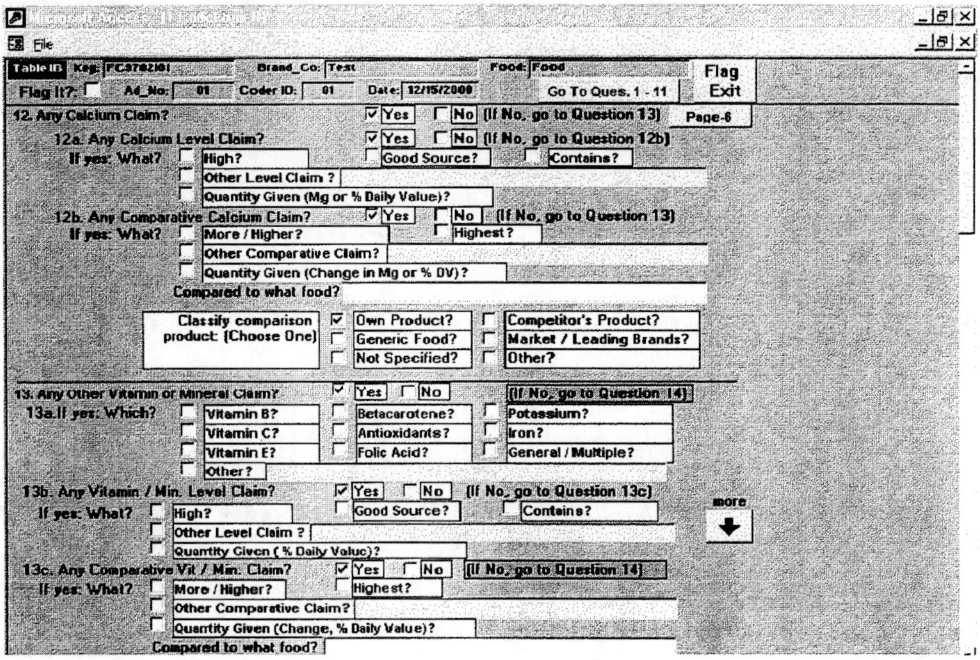
File
Table IB Key: FC9702101 Brand_Co: Test Food: Food Flag
Flag It? Ad_No: 01 Coder ID: 01 Date: 12/15/2000 Go To Ques. 1 - 11 Exit
12. Any Calcium Claim? Yes No (If No, go to Question 13) Page-6
12a. Any Calcium Level Claim? Yes No (If No, go to Question 12b)
If yes: What? High? Good Source? Contains?
Other Level Claim ?
Quantity Given (Mg or % Daily Value)?
12b. Any Comparative Calcium Claim? Yes No (If No, go to Question 13)
If yes: What? More / Higher? Highest?
Other Comparative Claim?
Quantity Given (Change in Mg or % DV)?
Compared to what food?
Classify comparison product: (Choose One) Own Product? Competitor's Product? Generic Food? Market / Leading Brands? Not Specified? Other?
13. Any Other Vitamin or Mineral Claim? Yes No (If No, go to Question 14)
13a. If yes: Which? Vitamin B? Betacarotene? Potassium? Vitamin C? Antioxidants? Iron? Vitamin E? Folic Acid? General / Multiple? Other?
13b. Any Vitamin / Min. Level Claim? Yes No (If No, go to Question 13c) more
If yes: What? High? Good Source? Contains?
Other Level Claim ?
Quantity Given (% Daily Value)?
13c. Any Comparative Vit / Min. Claim? Yes No (If No, go to Question 14)
If yes: What? More / Higher? Highest?
Other Comparative Claim?
Quantity Given (Change, % Daily Value)?
Compared to what food?

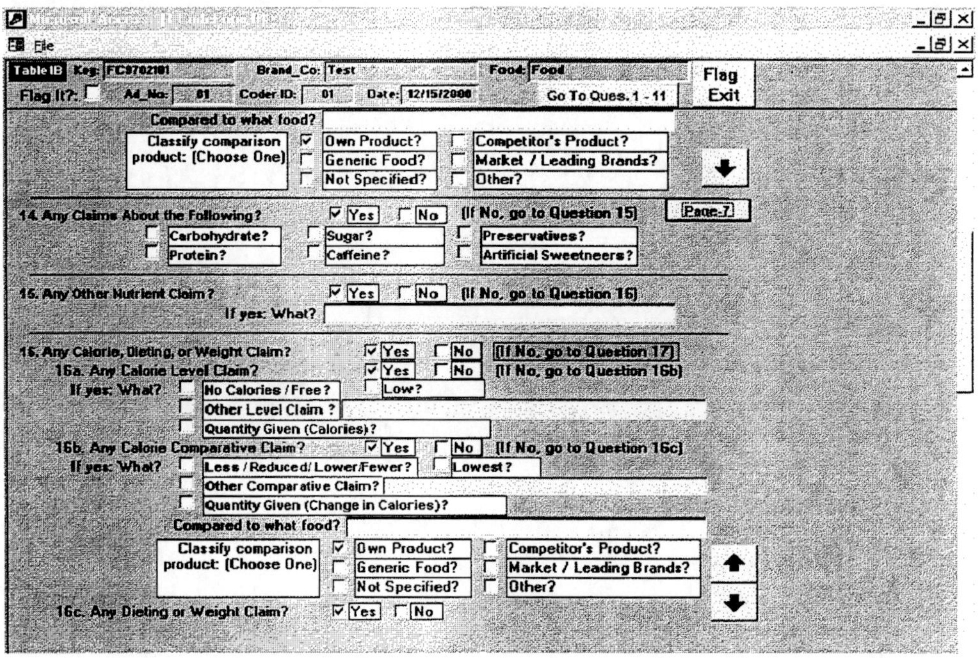
File
Table IB Key: FC9702101 Brand_Co: Test Food: Food Flag
Flag It? Ad_No: 01 Coder ID: 01 Date: 12/15/2000 Go To Ques. 1 - 11 Exit
Compared to what food?
Classify comparison product: (Choose One) Own Product? Competitor's Product? Generic Food? Market / Leading Brands? Not Specified? Other?
14. Any Claims About the Following? Yes No (If No, go to Question 15) Page-7
Carbohydrate? Sugar? Preservatives? Protein? Caffeine? Artificial Sweeteners?
15. Any Other Nutrient Claim? Yes No (If No, go to Question 16)
If yes: What?
16. Any Calorie, Dieting, or Weight Claim? Yes No (If No, go to Question 17)
16a. Any Calorie Level Claim? Yes No (If No, go to Question 16b)
If yes: What? No Calories / Free? Low?
Other Level Claim ?
Quantity Given (Calories)?
16b. Any Calorie Comparative Claim? Yes No (If No, go to Question 16c)
If yes: What? Less / Reduced / Lower/Fewer? Lowest?
Other Comparative Claim?
Quantity Given (Change in Calories)?
Compared to what food?
Classify comparison product: (Choose One) Own Product? Competitor's Product? Generic Food? Market / Leading Brands? Not Specified? Other?
16c. Any Dieting or Weight Claim? Yes No

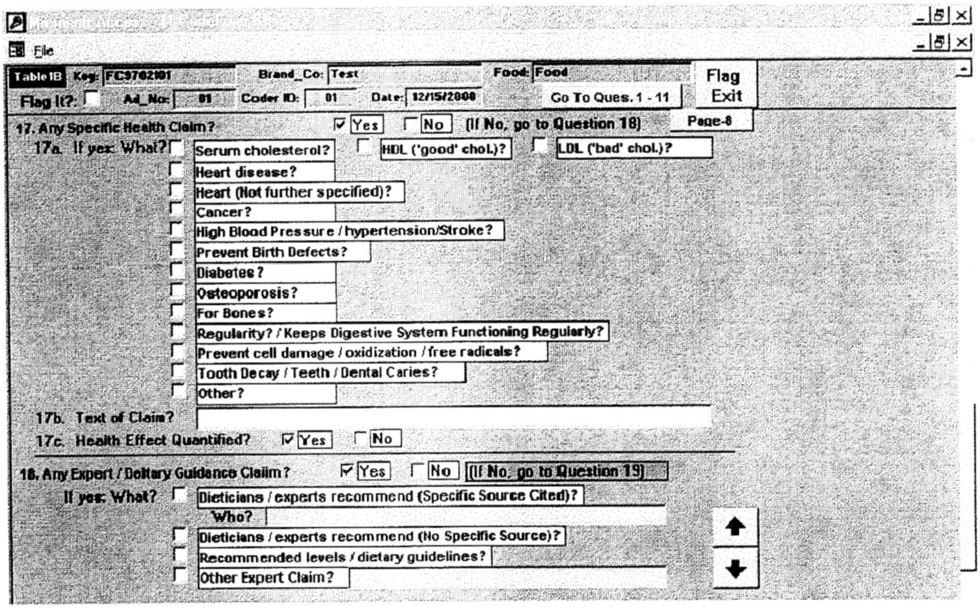
File
Table1B Key: FC9702101 Brand_Co: Test Food: Food
Flag It?: Ad_No: 01 Coder ID: 01 Date: 12/15/2000
Go To Ques. 1 - 11
Flag
Exit
Page-8
17. Any Specific Health Claim? Yes No (If No, go to Question 18)
17a. If yes: What?
Serum cholesterol?
HDL ('good' chol.)?
LDL ('bad' chol.)?
Heart disease?
Heart (Not further specified)?
Cancer?
High Blood Pressure / hypertension/Stroke?
Prevent Birth Defects?
Diabetes?
Osteoporosis?
For Bones?
Regularity? / Keeps Digestive System Functioning Regularly?
Prevent cell damage / oxidization / free radicals?
Tooth Decay / Teeth / Dental Caries?
Other?
17b. Text of Claim?
17c. Health Effect Quantified? Yes No
18. Any Expert / Dietary Guidance Claim? Yes No (If No, go to Question 19)
If yes: What?
Dieticians / experts recommend (Specific Source Cited)?
Who?
Dieticians / experts recommend (No Specific Source)?
Recommended levels / dietary guidelines?
Other Expert Claim?

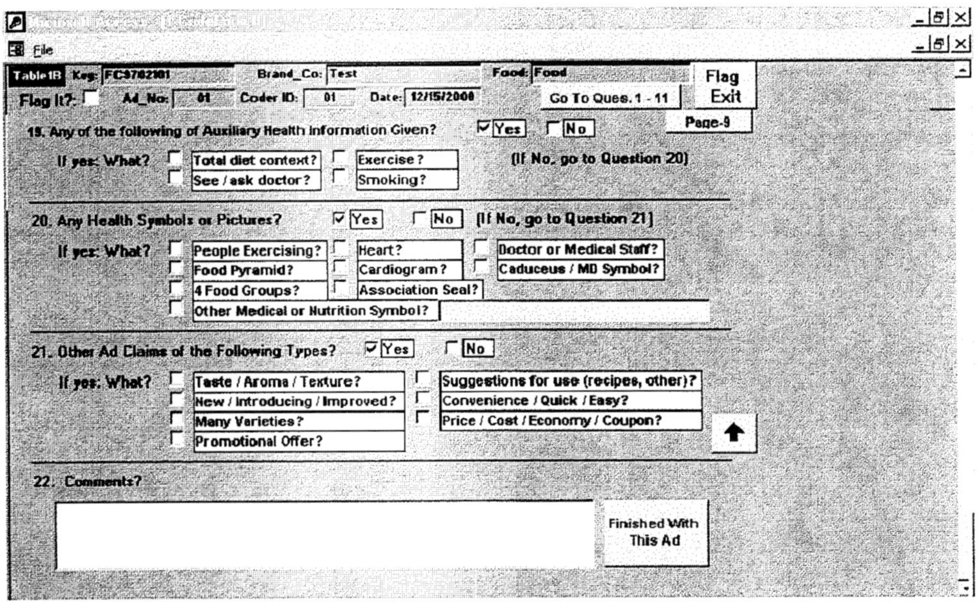
File
Table1B Key: FC9702101 Brand_Co: Test Food: Food
Flag It?: Ad_No: 01 Coder ID: 01 Date: 12/15/2000
Go To Ques. 1 - 11
Flag
Exit
Page-9
19. Any of the following of Auxiliary Health Information Given? Yes No
If yes: What?
Total diet context?
Exercise?
See / ask doctor?
Smoking?
(If No, go to Question 20)
20. Any Health Symbols or Pictures? Yes No (If No, go to Question 21)
If yes: What?
People Exercising?
Heart?
Doctor or Medical Staff?
Food Pyramid?
Cardiogram?
Caduceus / MD Symbol?
4 Food Groups?
Association Seal?
Other Medical or Nutrition Symbol?
21. Other Ad Claims of the Following Types? Yes No
If yes: What?
Taste / Aroma / Texture?
Suggestions for use (recipes, other)?
New / Introducing / Improved?
Convenience / Quick / Easy?
Many Varieties?
Price / Cost / Economy / Coupon?
Promotional Offer?
22. Comments?
Finished With This Ad

Appendix B.
Selected Data Tables

Table B-1 Percentage of Ads with *Nutrient Content Claims* By Type1[1]

Year	Any Lipid[2]	Fat[3]	Fat Level	Fat Compare	Saturated Fat	Sat. Level	Sat. Compare	Cholesterol	Chol. Level	Chol. Compare
1977	10.4	4.1	2.4	1.8	0.3	0.3	0.0	1.6	1.6	0.0
1978	12.0	3.6	1.9	2.0	0.9	0.9	0.0	3.8	3.8	0.5
1979	15.6	4.5	2.4	2.8	1.1	1.0	0.2	5.8	4.5	1.0
1980	12.6	3.8	1.8	2.1	1.1	0.8	0.3	4.6	4.1	0.7
1981	11.3	3.1	1.0	2.2	0.7	0.7	0.0	3.1	3.1	0.2
1982	11.9	5.5	2.9	2.9	0.7	0.7	0.0	6.4	6.2	0.7
1983	12.7	3.7	2.7	1.2	1.4	0.4	1.1	4.6	3.9	0.7
1984	13.7	4.9	2.5	3.1	0.8	0.3	0.5	4.4	3.3	1.3
1985	15.0	4.3	2.6	2.1	2.1	1.2	0.9	7.5	7.0	1.2
1986	12.9	4.6	3.3	1.8	2.0	1.6	1.0	7.0	6.5	1.1
1987	12.6	4.3	2.9	1.6	2.0	1.8	1.4	6.7	6.1	0.9
1988	17.6	8.9	6.3	3.9	3.9	3.3	1.5	12.6	12.1	1.5
1989	29.1	12.6	9.9	4.6	4.3	3.4	2.0	18.6	17.7	2.9
1990	31.2	18.4	12.4	9.2	7.7	7.7	0.9	21.6	19.9	4.5
1991	36.8	26.0	21.5	10.1	5.6	5.4	2.0	24.7	23.9	3.4
1992	31.1	26.4	23.1	7.8	4.1	3.3	3.7	16.9	15.9	3.5
1993	30.8	20.7	15.1	8.1	4.3	4.0	2.9	14.8	13.9	1.8
1994	31.4	25.8	20.3	7.8	2.7	2.3	2.1	10.1	10.1	0.2
1995	34.4	28.7	24.4	6.8	3.1	2.7	1.4	6.6	6.6	0.0
1996	40.2	35.7	30.6	10.5	2.7	2.5	0.6	6.0	6.0	0.4
1997	41.0	36.9	31.1	11.1	2.7	2.7	0.0	5.8	5.8	0.0

Table continued on next page.

Table B-1 — Continued

Year	Mono Fat	Mono Level	Mono Compare	Poly Fat	Poly Level	Poly Compare	Other Fat/Oil[4]	Corn Oil	Canola Oil	Olive Oil
1977	0.0	0.0	0.0	0.3	0.3	0.0	6.2	1.0	0.0	0.0
1978	0.0	0.0	0.0	0.9	0.6	0.6	9.4	1.1	0.0	0.2
1979	0.0	0.0	0.0	1.8	0.6	1.1	10.9	1.0	0.0	0.0
1980	0.0	0.0	0.0	0.7	0.5	0.2	7.0	0.5	0.0	0.2
1981	0.0	0.0	0.0	0.0	0.0	0.0	8.2	0.7	0.0	0.0
1982	0.0	0.0	0.0	1.0	1.0	0.0	7.9	2.1	0.0	0.0
1983	0.0	0.0	0.0	0.7	0.7	0.2	8.9	2.0	0.0	0.0
1984	0.0	0.0	0.0	1.0	0.8	0.8	10.7	3.6	0.0	0.2
1985	0.0	0.0	0.0	1.5	1.4	0.3	11.3	3.4	0.0	0.0
1986	0.2	0.2	0.0	0.5	0.5	0.0	7.0	1.1	0.0	0.0
1987	0.0	0.0	0.0	0.0	0.0	0.0	7.4	2.2	0.2	0.4
1988	0.2	0.2	0.0	0.9	0.9	0.0	7.8	3.0	0.7	1.1
1989	0.7	0.7	0.0	0.0	0.0	0.0	11.1	1.7	0.5	1.0
1990	0.2	0.2	0.0	0.0	0.0	0.0	9.2	2.3	0.6	0.0
1991	0.4	0.4	0.0	0.0	0.0	0.0	8.2	0.8	0.8	0.8
1992	0.0	0.0	0.0	1.6	1.6	0.0	7.4	0.6	0.2	0.2
1993	0.9	0.9	0.2	0.0	0.0	0.0	8.3	0.0	0.7	1.1
1994	0.8	0.8	0.0	0.0	0.0	0.0	10.1	0.0	0.6	0.4
1995	0.0	0.0	0.0	0.0	0.0	0.0	11.1	0.0	0.6	0.0
1996	0.0	0.0	0.0	0.0	0.0	0.0	8.8	1.0	1.0	1.4
1997	0.0	0.0	0.0	0.0	0.0	0.0	5.3	0.0	0.0	2.7

Table continued on next page.

Table B-1 — Continued

Year	Sodium	Sodium Level	Sodium Compare	Fiber	Fiber Level	Fiber Compare	Whole Grain	Calcium	Calcium Level	Calcium Compare
1977	1.9	1.9	0.0	2.6	2.6	0.8	1.0	2.6	2.4	0.5
1978	2.2	2.0	0.2	2.5	2.5	0.0	1.3	0.0	0.0	0.0
1979	2.1	1.9	0.3	3.6	3.6	1.1	0.6	0.6	0.6	0.5
1980	2.9	2.9	0.0	4.1	4.1	1.0	1.8	0.2	0.2	0.0
1981	3.3	3.3	0.2	6.2	6.2	1.0	3.1	0.0	0.0	0.0
1982	4.1	3.3	1.5	6.2	6.2	1.0	2.6	0.9	0.9	0.0
1983	8.0	6.4	2.0	5.5	5.5	0.9	2.0	0.7	0.4	0.4
1984	6.6	5.8	1.2	4.3	4.3	1.3	0.7	1.2	0.2	1.2
1985	7.0	6.7	0.6	5.8	5.8	2.6	2.0	2.0	1.4	1.5
1986	11.6	10.5	1.8	4.9	4.9	1.8	1.6	3.6	2.6	1.8
1987	11.0	9.0	4.7	7.6	7.2	2.2	4.0	4.0	3.4	1.8
1988	12.1	9.6	3.7	5.6	5.6	2.2	1.5	1.7	1.7	0.2
1989	11.1	9.4	3.7	9.5	9.5	4.1	2.2	3.7	3.7	2.4
1990	11.3	9.6	3.2	8.6	8.5	3.2	2.3	3.2	3.2	1.7
1991	13.3	9.9	6.0	5.4	5.4	1.0	2.4	2.4	2.4	0.2
1992	8.5	5.8	3.5	7.6	7.6	1.2	3.3	3.3	3.3	0.0
1993	8.8	5.8	3.8	8.3	8.3	0.0	5.2	1.6	1.6	0.7
1994	4.2	3.1	1.0	6.3	6.1	0.4	3.6	2.9	2.7	1.5
1995	7.6	4.9	3.1	9.6	9.6	0.0	4.5	7.6	7.0	1.0
1996	6.4	4.7	1.6	6.0	6.0	0.2	3.9	3.1	2.7	0.8
1997	6.0	5.1	1.0	6.5	6.5	0.2	1.0	3.9	3.9	0.5

Table continued on next page.

Table B-1 — Continued

Year	Vitamin	Vitamin Level	Vitamin Compare	Vitamin C	Vitamin E	Potassium	Folic Acid	General Vit/Min	Carbos	Sugar
1977	7.9	7.5	1.0	4.5	0.6	0.0	0.8	3.6	2.4	5.0
1978	7.4	6.5	1.6	2.2	0.2	0.0	0.3	3.9	0.5	7.6
1979	6.6	6.2	1.5	1.9	0.2	0.0	0.3	2.6	0.3	7.3
1980	3.6	3.4	0.5	1.6	0.5	0.3	0.3	1.1	0.0	5.1
1981	5.0	4.1	1.2	2.6	0.5	0.3	0.3	2.2	0.0	6.0
1982	6.5	5.5	2.4	3.1	0.0	0.9	0.0	2.6	0.7	6.5
1983	6.8	5.7	1.2	3.2	0.0	1.2	0.0	3.0	0.4	11.1
1984	7.6	7.6	0.0	2.5	0.0	0.5	0.2	2.8	0.8	10.0
1985	8.4	8.4	0.6	1.7	0.0	1.1	0.0	5.8	0.2	8.1
1986	7.0	6.9	0.3	2.6	0.2	2.1	0.2	1.1	0.0	8.8
1987	8.1	7.7	0.7	3.8	0.0	1.1	0.2	2.9	0.2	7.7
1988	6.1	6.1	0.0	2.2	0.0	0.7	0.0	2.8	0.9	7.2
1989	7.0	7.0	0.7	2.6	0.2	1.2	0.0	4.4	0.5	8.5
1990	10.2	9.0	2.6	4.7	0.0	1.3	0.0	4.7	1.5	8.5
1991	8.4	7.4	2.6	1.6	0.0	3.0	0.4	4.4	2.8	8.2
1992	8.4	6.6	2.1	1.9	0.0	1.7	0.0	4.7	4.7	4.5
1993	5.8	5.6	0.7	2.2	0.2	0.2	0.4	3.1	2.7	7.0
1994	5.9	5.9	0.0	3.8	1.3	0.8	1.0	2.5	0.8	5.0
1995	8.8	8.2	2.0	4.3	1.4	1.6	1.6	4.1	2.0	4.9
1996	7.6	7.6	0.6	4.5	0.8	1.2	1.2	2.3	1.4	4.9
1997	9.6	9.6	0.2	3.1	0.0	1.0	0.0	6.0	0.7	6.3

Table continued on next page.

Table B-1 — Continued

Year	No Preserves	Protein	Caffeine	Artificial Sweetener	Other Nutrient[5]	Calorie/ Diet[6]	Calorie Level	Calorie Compare	Diet	Any Nutrient[7]
1977	3.2	6.0	0.3	0.8	1.6	7.8	5.2	1.3	4.7	28.5
1978	3.6	3.9	2.4	0.6	2.5	6.0	3.8	2.5	4.4	33.5
1979	2.4	4.1	2.3	0.3	0.8	9.1	5.5	3.7	4.5	36.3
1980	4.7	2.8	3.4	1.6	1.5	7.7	5.4	2.9	3.3	33.8
1981	6.4	3.1	1.7	0.5	2.6	7.6	3.4	3.8	5.0	35.9
1982	6.7	4.5	1.4	1.2	3.1	9.5	4.5	4.8	5.3	37.0
1983	6.8	5.7	2.7	2.5	1.6	12.3	6.6	5.2	6.1	43.7
1984	3.9	3.6	3.5	6.6	0.5	13.5	8.9	6.1	6.3	43.8
1985	5.5	3.4	2.7	6.0	0.5	16.5	9.8	6.0	6.1	46.3
1986	6.9	1.0	2.1	4.4	0.8	14.9	9.0	6.4	4.9	46.2
1987	5.2	2.9	5.6	4.0	1.1	18.2	10.3	8.3	5.9	52.3
1988	0.9	3.7	3.0	3.2	2.0	16.7	9.1	7.2	4.1	44.2
1989	4.8	2.0	3.9	4.1	1.0	22.1	11.6	11.2	4.9	62.2
1990	5.8	3.6	2.6	4.7	1.7	20.9	9.4	10.3	5.5	55.8
1991	5.6	3.6	2.4	2.6	2.2	22.5	14.5	10.3	5.0	56.3
1992	3.7	5.6	2.1	1.0	1.2	13.6	7.4	6.0	2.1	47.2
1993	3.6	2.5	1.3	3.4	0.7	17.1	11.0	6.5	1.3	55.5
1994	3.6	0.8	1.0	0.8	0.4	10.5	6.7	3.1	3.4	47.8
1995	2.0	2.7	0.8	1.0	4.5	8.6	6.3	1.2	2.0	49.6
1996	0.8	2.1	2.9	2.1	0.8	13.8	9.9	2.7	2.9	56.1
1997	1.9	2.7	1.7	1.2	2.4	12.0	8.7	3.4	2.7	58.1

Table continued on next page.

Table B-1 — Continued

Notes. [1] Percentage of ads with a claim for the listed nutrient. See Chapter 4 for more specific definitions. *Level* claim refer to the absolute amount of the nutrient, as in *low fat*, and *comparative* claims compare the nutrient amount to something else (even if unstated), as in *less fat.* Level claims and comparative claims are both included in the overall category for the nutrient, *e.g., fat level* and *fat comparative* claims are both in the *fat* claim category. The level and comparative categories do not necessarily add to the overall category, because an ad can have claims from both.

[2] The *any lipid* category includes any advertisement with claims about fat, saturated fat, cholesterol, polyunsaturated fat, monounsaturated fat, type of oil used, or any other fat or oil claim, such as *baked not fried.*

[3] The *fat* category includes claims about fat that are not further qualified. In particular, it does not include saturated fat or other type of fat claims.

[4] The *other fat/oil* category includes type of oil claims or any other fat-related claims not in our primary categories.

[5] The *other nutrient* category includes advertisements with a nutrient-related claim not explicitly reflected in our coding system.

[6] The *calorie/diet* category includes advertisements with claims about calories, dieting, or weight control. The *calorie level*, *calorie comparative*, and *diet* categories are all subcategories of the *calorie/diet* category.

[7] The *any nutrient* category includes advertisements with any of the nutrient-related claims reflected in our coding system.

Table B-2 Percentage of Ads with *General Nutrition Claims* by Type[1]

Year	Any General	Healthy	Smart/ Right	Good For You	Nutritious	Enriched/ Fortified	Wholesome	Light	Lean	Guilt Free
1977	49.4	1.6	2.1	6.0	7.5	6.6	1.9	6.3	1.8	0.6
1978	52.1	1.4	3.5	6.1	6.5	5.2	2.4	7.1	2.0	1.4
1979	52.8	1.6	1.1	6.0	6.0	3.7	2.9	14.6	2.8	1.6
1980	55.1	1.1	2.3	5.7	5.4	1.3	1.5	13.7	2.6	2.0
1981	56.4	1.4	1.0	6.4	4.6	2.6	3.3	11.3	1.5	0.5
1982	62.0	4.6	3.4	10.8	5.7	0.9	4.6	13.6	3.8	2.1
1983	71.1	3.4	3.2	11.6	9.3	3.2	3.9	15.2	4.1	0.9
1984	69.6	4.8	3.1	12.2	5.6	3.0	5.3	11.7	3.5	1.3
1985	67.9	3.7	3.8	11.8	4.3	3.5	4.7	16.2	2.9	1.8
1986	65.2	3.8	4.4	7.0	5.6	2.6	4.6	16.8	3.1	1.3
1987	66.1	3.2	4.3	7.6	8.5	3.1	3.4	16.8	3.8	1.8
1988	67.2	4.6	4.3	7.2	7.8	2.6	0.7	15.0	2.0	2.0
1989	68.0	10.6	5.8	11.2	8.5	4.1	2.7	14.0	1.9	1.0
1990	66.2	12.6	7.9	9.0	12.0	5.8	4.5	16.2	1.7	1.3
1991	66.8	10.1	4.6	14.7	8.5	4.4	3.8	17.5	3.2	3.6
1992	58.1	9.5	4.1	12.2	9.5	4.5	1.4	13.0	2.5	2.9
1993	70.1	10.3	5.6	10.1	5.6	3.6	3.8	22.0	2.9	1.3
1994	59.1	6.9	4.6	5.9	8.8	2.1	3.6	12.8	4.0	0.8
1995	62.5	7.4	2.0	6.6	15.4	2.0	4.1	12.3	1.6	1.0
1996	55.2	9.0	3.9	6.0	10.5	4.3	1.4	10.1	2.9	2.3
1997	55.9	8.9	0.7	5.3	10.8	3.6	1.2	10.1	3.6	3.1

Table continued on next page.

Table B-2 — Continued

Year	Fresh	Natural/ Real	Energy	Young/ Fit	Other General[2]	Smart/ Good[3]	Nutritious/ Wholesome[4]	Light/ Lean[5]	Core General[6]
1977	12.6	27.2	0.5	1.5	0.0	8.3	13.6	8.1	19.6
1978	15.6	30.6	1.1	3.6	0.0	12.0	12.1	8.8	22.7
1979	15.4	27.6	0.3	2.8	0.0	9.9	10.9	16.0	26.9
1980	16.5	29.4	0.0	3.3	0.0	9.8	7.5	14.7	24.7
1981	15.6	35.2	0.0	1.7	0.0	8.6	9.1	12.2	22.7
1982	21.2	37.5	0.5	1.7	0.0	16.9	10.3	16.7	30.5
1983	17.8	44.6	1.4	4.6	0.2	17.3	15.0	18.7	36.5
1984	16.6	47.2	0.2	5.8	0.2	19.6	13.7	14.8	36.2
1985	16.5	42.9	0.8	2.3	0.0	16.2	12.2	18.8	35.7
1986	17.5	39.5	0.2	2.9	0.0	13.7	11.9	19.1	33.0
1987	17.1	39.5	0.5	3.8	0.0	13.9	11.7	19.6	34.2
1988	14.3	42.3	0.7	3.5	0.2	14.5	10.4	16.1	30.1
1989	16.4	43.4	0.2	4.8	0.0	20.6	13.5	15.2	35.3
1990	18.2	33.3	2.6	5.1	0.2	22.7	20.5	16.7	38.7
1991	10.7	34.4	2.8	4.4	0.0	23.3	15.9	20.1	39.6
1992	12.8	36.5	2.9	5.8	0.0	20.8	12.6	15.3	32.0
1993	12.4	36.0	2.7	3.6	0.2	19.8	12.4	24.0	43.1
1994	12.4	32.9	0.8	3.8	0.0	15.3	14.0	16.6	34.0
1995	9.8	30.5	1.0	3.7	0.0	14.6	21.1	13.9	34.0
1996	5.5	27.5	1.6	2.9	0.2	16.4	15.8	12.9	30.6
1997	8.4	26.7	1.9	4.3	0.0	12.5	14.0	13.3	31.6

Table continued on next page.

Table B-2 — Continued

Notes. [1] Percentage of ads with a claim from the listed category of general nutrition claims. For specific definitions, see Chapter 4.

[2] Other general claims includes any general nutrition claim not specifically listed.

[3] Includes all claims from the *smart/right choice*, *good for you*, and *health/healthy* categories.

[4] Includes all claims from the *nutritious, wholesome*, and *enriched/fortified* categories.

[5] Includes all claims from the *light* and *lean* categories.

[6] *Core general claims* includes all claims from the *healthy/health*, *smart/right choice*, *good/better for you*, *nutritious/nutrients*, *wholesome*, *enriched/fortified*, *light/lighter*, *lean/leaner*, and *youth/fitness/well-being* claims. Thus, the category excludes *fresh*, *natural/no artificial/real/pure*, *energy*, *guilt free/no guilt/cheating* claims, as well as the miscellaneous *other* category.

Table B-3 Percentage of Ads with *Health Claims* by Type[1]

Year	Any Health[2]	Disease[3]	Disease/ Affiliated[4]	Other Health[5]	Heart Disease	Heart (NFS)[6]	Serum Chol.	Cancer	Blood Pressure	Birth Defects
1977	1.8	0.0	0.3	1.5	0.0	0.0	0.3	0.0	0.0	0.0
1978	3.5	0.8	1.4	2.0	0.8	0.3	1.4	0.0	0.0	0.0
1979	3.4	0.5	1.1	2.3	0.5	0.2	1.1	0.0	0.2	0.0
1980	2.0	0.5	1.1	0.8	0.2	0.0	0.8	0.3	0.2	0.0
1981	1.9	0.0	0.0	1.9	0.0	0.0	0.0	0.0	0.0	0.0
1982	2.9	0.0	0.3	2.6	0.0	0.2	0.0	0.0	0.2	0.0
1983	2.7	0.4	1.8	0.9	0.2	0.7	1.4	0.0	0.7	0.0
1984	3.3	0.8	2.0	1.3	0.3	1.0	0.7	0.5	1.0	0.0
1985	3.8	1.4	3.1	1.1	0.3	1.2	1.4	0.8	0.2	0.0
1986	3.9	1.0	2.3	2.5	0.5	0.7	0.8	0.3	0.7	0.0
1987	4.5	1.8	3.2	2.0	0.7	1.4	1.3	0.9	0.4	0.0
1988	6.5	2.8	5.6	1.3	2.8	1.7	4.6	0.2	0.2	0.0
1989	11.1	3.4	8.7	3.6	2.9	3.2	6.6	0.5	0.3	0.0
1990	9.8	3.6	7.7	2.8	2.6	3.4	5.3	0.6	0.0	0.0
1991	7.0	2.4	5.2	2.0	2.0	2.6	2.4	0.4	0.0	0.0
1992	2.5	0.2	1.7	0.8	0.2	1.7	0.0	0.0	0.0	0.0
1993	2.2	0.4	0.7	1.8	0.4	0.2	0.2	0.0	0.0	0.0
1994	2.5	0.8	0.8	2.5	0.0	0.0	0.0	0.8	0.0	0.8
1995	5.5	2.1	2.7	3.3	1.4	1.6	0.4	0.4	1.2	0.2
1996	2.9	1.8	2.5	0.4	1.0	0.6	0.8	1.0	1.0	0.8
1997	8.2	4.6	6.3	1.9	1.9	2.2	1.0	2.2	0.2	0.0

Table continued on next page.

Table B-3 — Continued

Year	Osteoporosis	Bones	Diabetes	Regularity/ Digestion	Cell Damage	Tooth Decay	Other[7]	Any Heart[8]	Any Bone[9]
1977	0.0	0.5	0.0	0.3	0.0	0.3	0.8	0.3	0.5
1978	0.0	0.0	0.0	0.0	0.0	0.0	2.0	1.4	0.0
1979	0.0	0.0	0.0	0.0	0.0	0.2	2.1	1.1	0.0
1980	0.0	0.0	0.0	0.0	0.0	0.3	0.5	0.8	0.0
1981	0.0	0.0	0.0	0.0	0.0	0.0	1.9	0.0	0.0
1982	0.0	0.0	0.0	1.5	0.0	0.0	1.0	0.2	0.0
1983	0.0	0.0	0.2	0.0	0.0	0.2	0.7	1.6	0.0
1984	0.0	0.0	0.0	0.0	0.0	1.0	0.7	1.3	0.0
1985	0.3	0.9	0.0	0.0	0.0	0.2	0.5	2.3	0.9
1986	0.2	0.8	0.0	0.2	0.0	0.3	1.5	1.3	0.8
1987	0.2	0.7	0.0	0.0	0.0	0.9	0.7	2.0	0.7
1988	0.0	0.4	0.0	0.2	0.0	0.4	0.4	5.6	0.4
1989	0.2	1.2	0.2	0.3	0.0	0.9	1.9	8.2	1.2
1990	0.4	2.1	0.0	0.0	0.0	0.6	0.8	6.8	2.1
1991	0.0	0.2	0.0	0.0	0.0	0.0	2.0	4.6	0.2
1992	0.0	0.0	0.0	0.4	0.0	0.0	0.4	1.7	0.0
1993	0.0	0.9	0.0	0.4	0.0	0.0	0.4	0.7	0.9
1994	0.0	1.3	0.0	0.0	0.0	0.0	1.3	0.0	1.3
1995	0.4	2.1	0.0	0.0	0.0	0.6	2.0	2.0	2.1
1996	0.4	0.0	0.0	0.0	0.0	0.0	0.4	2.1	0.4
1997	0.5	0.5	0.0	0.0	0.0	0.0	1.4	3.4	1.0

Table continued on next page.

Table B-3 — Continued

Notes. [1] Percentage of ads with a claim from the listed category of health claims. For definitions, see Chapter 5.

[2] Percentage of ads with any type of health claim, that is, any statement or term referring to specific health effects of nutrients or foods.

[3] *Disease claims* specifically refer to diseases, such as "saturated fat has been linked to heart disease."

[4] *Disease and affiliated claims* includes disease claims and claims closely affiliated with diseases, specifically serum cholesterol claims, general *heart* claims, such as "heart smart," and high blood pressure claims.

[5] *Other health claims* includes all health claims that are not disease or affiliated claims. These are often structure-function claims in FDA terminology.

[6] NFS is not further specified.

[7] Includes all health claims other than the specific categories listed.

[8] Combines heart disease, heart (NFS), and the serum cholesterol claims.

[9] Combines osteop orosis and bone claims.

APPENDIX C.

FTC FOOD ADVERTISING CASES[1]

January 1977 - April 2002

Interstate Bakeries Corp., C-4043 (April 16, 2002) (consent) (alleged unsubstantiated claims that its Wonder Bread containing added calcium could improve children's brain function and memory)

Abbott Laboratories, C-3745 (May 30, 1997) (consent) (many doctors recommend Ensure nutritional beverages as a meal supplement and replacement for healthy adults)

Gerber Products Co., C-3744 (May 27, 1997) (consent) (4 out of 5 pediatricians recommend Gerber baby food)

Pizzeria Uno Corp., C-3730 (Apr. 4, 1997) (consent) (fat content of thin crust pizzas)

Conopco, Inc., C-3706 (Jan. 23, 1997) (consent) ("Get Heart Smart" campaign for Promise margarines and spreads)

Mrs. Fields Cookies, Inc., C-3657 (May 13, 1996) (consent) (fat content of new line of fresh-baked cookies)

Mama Tish's Italian Specialties, Inc., C-3644 (Mar. 19, 1996) (consent) (calorie content of flavored ice-cup desserts)

The Dannon Co., C-3643 (Mar. 18, 1996) (consent and $150,000 in disgorgement) (fat and calorie content of Pure Indulgence frozen yogurt)

Eggland's Best, Inc., US v., No. 96 CV-1983 (E.D. Pa. Mar. 12, 1996) (stipulated permanent injunction and $100,000 civil penalty - violation of previous order, below) (effect of its eggs on serum cholesterol)

Good News Products, Inc., C-3642 (Feb. 22, 1996) (consent) (fat content of its eggs; effect of its eggs on risk factors for heart disease and blood cholesterol levels)

The Eskimo Pie Corp., C-3597 (Aug. 11, 1995) (consent) (calorie content of Sugar Freedom line of frozen dessert products, and American Diabetes Association endorsement)

Haagen Dazs Co., C-3582 (June 2, 1995) (consent) (fat content of frozen yogurt line of products)

Stouffer Foods Corp., D. 9250 (Sept. 26, 1994) (Commission Decision) (sodium content of Lean Cuisine frozen entrees)

[1] This list contains all FTC food advertising cases during the years listed. The type of claim at issue in the case is noted in the parentheses.

Eggland's Best, Inc., C-3520 (Aug. 15, 1994) (consent) (effect of its eggs on serum cholesterol)

Presto Food Products, Inc., C-3480 (Feb. 23, 1994) (consent) (fat content of Mocha Mix non-dairy creamer products)

Gracewood Fruit Co., 116 F.T.C. 1262 (1993) (consent) (health benefits of grapefruit, such as reduction of risk of cancer, stroke, heart attack; and reduction of serum cholesterol)

The Clorox Co., 116 F.T.C. 346 (1993) (consent) (fat content of Take Heart salad dressing)

The Isalay Klondike Co., 116 F.T.C. 74 (1993) (consent) (fat and calorie content of Klondike Lite frozen dessert bars)

Pompeian, Inc., 115 F.T.C. 933 (1992) (consent) (effect of olive oil on blood cholesterol and heart health in comparison with vegetable oil)

Campbell Soup Co., 115 F.T.C. 788 (1992) (consent) (effect of fat and cholesterol content of soups on risk of heart disease; high sodium content not disclosed)

Pacific Rice Products, Inc., 115 F.T.C. 763 (1992) (consent) (effect of Vita-Fiber Rice Bran on blood cholesterol and risk of heart disease)

Bertolli USA, Inc., 115 F.T.C. 774 (1992) (consent) (effect of olive oil on blood pressure and blood sugar; and effect on cholesterol in comparison to other oils)

Nestle Food Co., 115 F.T.C. 67 (1992) (consent) (fat content of Carnation Coffee-mate Liquid)

The Perrier Group of America, Inc., 114 F.T.C. 486 (1991) (consent) (sparkling mineral water is not processed or filtered before bottling)

Kraft Inc., 114 F.T.C. 40 (1991), aff'd, 970 F.2d 311 (7th Cir. 1992), cert. denied, 113 S. Ct. 1254 (1993) (calcium content of Kraft Singles cheese slices compared to milk and compared to most imitation cheese slices)

CPC International, Inc., 114 F.T.C. 1 (1991) (consent) (effect of Mazola Corn Oil and Mazola Margarine on serum cholesterol levels)

Meadow Fresh Farms Inc. (Roy and Larry Brog), 108 F.T.C. 18 (1986) (consent) (effect of dry milk substitute on cardiovascular disease)

Estee Corp., 102 F.T.C. 1804 (1983) (consent and $25,000 to American Diabetes Association for research) (calorie content of special health-related foods appropriate for diabetics)

Standard Brands, Inc., C-3060 (March 17, 1981) (consent) (twice as many doctors use and recommend Fleischmann's Margarine)

California Milk Producers Advisory Board, 8988 (September 21, 1979) (final order dismissing complaint) ("Every body needs milk" ad campaign)

ITT Continental Baking Company, Inc., C-2989 (August 24, 1979) (consent) (dietary fiber in Fresh Horizons bread, and other such products)

National Commission on Egg Nutrition, 8987 (November 17, 1978) (modifying order based on 7th Cir. decision) (effect of eating eggs on likelihood of heart disease and arteriosclerosis)

California and Hawaiian Sugar Co., C-2858 (January 6, 1977) (consent) (C&H sugar is different from and superior to other sugar brands)

Ad Agencies

Campbell Mithun LLC C-4043 (April 16, 2002) (consent) (ad agency for Interstate Bakeries Corp.[IBC], advertiser of Wonder Bread)

Grey Advertising, Inc., C-3691 (Oct. 30, 1996) (consent) (ad agency for The Dannon Co., advertiser of Pure Indulgence frozen yogurt)

N.W. Ayer & Son, Inc., C-3660 (May 31, 1996) (consent) (ad agency for Eggland's Best, Inc., advertiser of Eggland's Best eggs)

BBDO Worldwide, Inc., C-3637 (Jan. 24, 1996) (consent) (ad agency for Haagen Dazs, advertiser of frozen yogurt line of products)

Ted Bates & Company, Inc., C-3059 (March 17, 1981) (consent) (ad agency for Standard Brands, Inc., advertiser of Fleischmann's Margarine)

Foote, Cone & Belding/Honing, Inc., C-2858 (January 6, 1977) (consent) (ad agency for California and Hawaiian Sugar Company, advertiser of C&H sugar)

REFERENCES

Bagwell, Kyle, "The Economics of Advertising, Introduction," in *The Economics of Advertising*, Kyle Bagwell, Editor, Northampton, MA: Edward Elgar, 2001.

Beales, J. Howard and Timothy J. Muris, *State and Federal Regulation of National Advertising*, AEI Studies in Regulation and Federalism, Washington, DC: The AEI Press, 1993.

Becker, Gary S. and Kevin M. Murphy, "A Simple Theory of Advertising As a Good or Bad," *Quarterly Journal of Economics*, 58 November 1993.

Benham, Lee, "The Effects of Advertising on the Price of Eyeglasses," *Journal of Law and Economics*, 15 October 1972.

Bond, Ronald S., John E. Kwoka, John J. Phelan, and Ira Taylor, *Effects of Restrictions on Advertising and Commercial Practice in the Professions: The Case of Optometry*, Bureau of Economics Staff Report, Federal Trade Commission, Washington, DC, September 1980.

Butters, Gerard R., "Equilibrium Distribution of Sales and Advertising Prices," *The Review of Economic Studies*, 44 October 1977.

Cady, John, "An Estimate of the Price Effects of Restrictions on Drug Price Advertising," *Economic Inquiry*, 14 December 1976.

Calfee, John E. and Debra Jones Ringold, "The 70% Majority: Enduring Consumer Beliefs About Advertising," *Journal of Public Policy & Marketing*, 13 Fall 1994.

Crespi, John M., "Promotion Checkoffs, Why So Controversial? The Evolution of Generic Advertising Battles," National Institute for Commodity Promotion Research & Evaluation Research Bulletin No. 2001-04, September 2001.

Derby, Brenda M. and Alan S. Levy, "Do Food Labels Work? Gauging the Effectiveness of Food Labels Pre- and Post-NLEA," in *Handbook of Marketing and Society*, Paul N. Bloom and Gregory T. Gundlach, Editors, Thousand Oaks, CA: Sage Publications, 2000.

Dixon, William D., *Report of the Presiding Officer, Proposed Trade Regulation Rule: Food Advertising*, Federal Trade Commission, Washington, D.C., 1978.

Dowling, Grahame R., "Information Content in U.S. and Australian Television Advertising," *Journal of Marketing*, 44 Fall 1980.

Ehrlich, Isaac and Lawrence Fisher, "The Derived Demand for Advertising: A Theoretical and Empirical Investigation," *The American Economic Review*, 72 June 1982.

Glickman, Secretary of Agriculture, v. Wileman Brothers & Elliott, Inc. et al., 521 U.S. (1997).

Glazer, Amihai, "Advertising, Information, and Prices – A Case Study," *Economic Inquiry*, 19 October 1981.

Greene, William H., *Econometric Analysis*, 3rd ed., Upper Saddle River, NJ: Prentice Hall, 1997.

Grossman, Gene M. and Carl Shapiro, "Informative Advertising With Differentiated Products," *Review of Economic Studies*, 51, January1984.

Grossman, Sanford J., "The Informational Role of Warranties and Private Disclosure About Product Quality," *Journal of Law and Economics*, 24, December 1981.

Heimbach, James T., "The Growing Impact of Sodium Labeling of Foods," *Food Technology*, 40 December 1986.

Hickman, Beth Wallace, Gail E. Gates, and Richard P. Dowdy, "Nutrition Claims in Advertising: A Study of Four Women's Magazines," *Journal of Nutrition Education,* 225 September/October 1993.

Ippolito, Pauline M., "How Government Policies Shape the Food and Nutrition Information Environment," *Food Policy,* 24 April/June 1999.

_____, "Bonding and Nonbonding Signals of Product Quality," *The Journal of Business*, 63, January 1990.

Ippolito, Pauline M. and Alan D. Mathios, *Health Claims in Advertising and Labeling: A Study of the Cereal Market*, Bureau of Economics Staff Report, Federal Trade Commission, Washington, D.C., August 1989.

_____, "Information, advertising and health choices: a study of the cereal market," *RAND Journal of Economics,* 21 Autumn 1990.

_____, "New Food Labeling Regulations and the Flow of Nutrition Information to Consumers," *Journal of Public Policy and Marketing*, 12 Fall 1993.

_____, "Information and Advertising: The Case of Fat Consumption in the United States," *American Economic Review*, 85 May 1995.

_____, *Information and Advertising Policy, A Study of Fat and Cholesterol Consumption in the United States, 1977-1990*, Bureau of Economics Staff Report, Federal Trade Commission, September 1996.

Kassarjian, Harold H., "Content Analysis in Consumer Research," *Journal of Consumer Research*, 4 June 1977.

Kihlstrom, Richard E. and Michael H. Riordan, "Advertising As a Signal," *Journal of Political Economy*, 92 June 1984.

Klein, Benjamin and Keith B. Leffler, "The Role of Market Forces in Assuring Contractual Performance," *Journal of Political Economy*, 89 August 1981.

Laband, David N., "The Durability of Informational Signals and the Content of Advertising," *Journal of Advertising*, 18, Spring 1989.

Leading National Advertisers, Inc., *Ad $ Summary*, New York: Leading National Advertisers, Inc., various years.

Lord, John B., Joseph O. Eastlack, and John L. Stanton, "Health Claims in Food Advertising: Is There a Bandwagon Effect?" *Journal of Advertising Research*, 27 April/May 1987.

_____, "The Bandwagon Isn't Rolling ...Yet," *Journal of Advertising Research*, 28 April/May 1988.

Maurizi, Alex R., "The Effect of Laws Against Price Advertising: The Case of Retail Gasoline," *Western Economic Journal*, 10 September 1972.

Milyo, Jeffrey and Joel Waldfogel, "The Effect of Price Advertising on Prices: Evidence in the Wake of *44 Liquormart*," *The American Economic Review*, 89 December 1999.

Moorman, Christine, "Market-Level Effects of Information: Competitive Responses and Consumer Dynamics," *35 Journal of Marketing Research*, 1998.

National Research Council, Committee on Diet and Health, *Diet and Health, Implications for Reducing Chronic Disease Risk*, National Academy Press, Washington, DC, 1989.

Nelson, Phillip, "Information and Consumer Behavior," *Journal of Political Economy*, 78 March-April 1970.

_____, "Advertising as Information," *Journal of Political Economy*, 82 July-August 1974.

Pappalardo, Janis Kohanski and Debra Jones Ringold, "Regulating Commercial Speech in a Dynamic Environment: Forty Years of Margarine and Oil Advertising Before the NLEA," *Journal of Public Policy and Marketing*, 19 Spring 2000.

Parker, Philip M., "*Sweet Lemons*: Illusory Quality, Self-Deceivers, Advertising, and Price," 32 *Journal of Marketing Research*, August 1995.

Pratt, Charlotte A. and Cornelius B. Pratt, "Comparative Content Analysis of Food and Nutrition Advertisements in *Ebony, Essence,* and *Ladies Home Journal*," *Journal of Nutrition Education*, 27 January/February 1995.

Resnik, A. J. and B. L. Stern, "An Analysis of Information Content in Television Advertising," *Journal of Marketing*, 41 January 1977.

Ringold, Debra Jones and John E. Calfee, "What Can We Learn From the Informational Content of Cigarette Advertising? A Reply and Further Analysis," *Journal of Public Policy and Marketing*, 8 1989.

Stigler, George J., "The Economics of Information," *Journal of Political Economy*, 69 June 1961.

Teisl, Mario F., Alan S. Levy, and Brenda M. Derby, "The Effects of Education and Information Source on Consumer Awareness of Diet-Disease Relationships," 18 *Journal of Public Policy & Marketing*, Fall 1999.

Telser, Lester G., "Advertising and Cigarettes," *Journal of Political Economy*, 70 October 1962.

Weimer, Jon, "Accelerating the Trend Towards Healthy Eating," in *America's Eating Habits, Change and Consequence*, Elizabeth Frazao Editor, U.S. Department of Agriculture, Economic Research Service, Agr. Info. Bul. No. 750, 1999.

INDEX

H

I

J

K

L

M

N

O

P

R

S